HOW TO PASS

NUMERICAL REASONING TESTS

HOW TO PASS

NUMERICAL REASONING TESTS

A step-by-step guide to learning key numeracy skills

Revised Edition

INTERMEDIATE LEVEL

HEIDI SMITH

KOGAN PAGE

London and Philadelphia

While the author has made every effort to ensure that the content of this book is accurate, please note that occasional errors can occur in books of this kind. If you suspect that an error has been made in any of the tests included in this book, please inform the publisher at the address below so that it can be corrected at the next reprint.

First published in Great Britain and the United States in 2003 by Kogan Page Limited
Revised edition 2006
Reprinted 2007, 2008

120 Pentonville Road
London N1 9JN
United Kingdom
www.koganpage.com

525 South 4th Street, #241
Philadelphia PA 19147
USA

ISBN-10 0 7494 4796 6
ISBN-13 978 0 7494 4796 0

British Library Cataloguing-in-Publication Data

A CIP record for this book is available from the British Library.

Library of Congress Cataloging-in-Publication Data

Smith, Heidi.
How to pass numerical reasoning tests: a step-by-step guide to learning key numeracy skills/Heidi Smith. -- Rev. ed.
p. c.m.
ISB 0-7494-4796-6
1. Mathematics–Examinations, questions, etc. I. Title.
QA43.S654 2006
510.76--dc22

2006019920

Typeset by Jean Cussons Typesetting, Diss, Norfolk
Printed and bound in Great Britain by MPG Books Ltd, Bodmin, Cornwall

This book is dedicated to Dr J V Armitage

Contents

Preface to revised edition

The success of the First Edition of *How to Pass Numerical Reasoning Tests* demonstrates that there are many people who, despite a high level of experience and competence in other areas of their professional life, are not confident about passing tests that involve numbers. However, what is not commonly known is that with practice it is possible to improve your numerical reasoning score dramatically.

This book was designed with the adult test-taker in mind – someone who is looking for some extra practice material prior to taking a test. Topics covered include:

- **percentages** – 'what are the three parts I have to know in order to work out the answer?'
- **decimals** – 'what was that quick way to multiply them?'
- **workrates** – 'wasn't there something about adding the total time together and dividing by something?'
- **prime numbers and multiples** – 'I remember I can use these to help me, but can't quite remember how'
- **interest rates** – 'what were the two different types and how do I use each of them?'.

This workbook reminds the non-mathematician what these terms are and how to work with them quickly.

The First Edition of *How to Pass Numerical Reasoning Tests* has helped thousands of applicants prepare for their test. This Revised Edition will help many more. Better preparation for psychometric testing leads to increased confidence during the initial assessment and this, undoubtedly, gives you a better chance of getting that job.

In addition to the friends and colleagues who reviewed the First Edition of *How to Pass Numerical Reasoning Tests*, I would also like to say thank you to Cheryl Hornbaker for her invaluable advice in preparing this revised edition.

Introduction

Numerical reasoning tests in context

Standardized testing is becoming more popular as a means to assess candidates at an early stage of the job application process, particularly in organizations where the demand for jobs is high. If you have picked up this book and started to read this chapter, the chances are that you are facing a numerical reasoning test in the very near future. It is likely that you know that you are perfectly capable of doing the job for which you have applied, and that the only obstacle between you and the next round of interviews is a set of tests that includes a numerical reasoning section. It is possible that you are dreading it. Not many people like being tested under pressure, so it is unsurprising that you are not looking forward to the experience. Take heart though. The good news is that the numerical knowledge you need to do well in these tests is the maths you learnt in school. You need now a quick refresher course and the application of some logical thinking. With practice and commitment to drilling in mental arithmetic you can improve your score. The numerical reasoning section of an aptitude test is the section where many people find they can improve their score, so it's worth dedicating a decent amount of test preparation time to this area.

Purpose of this workbook

This is a self-study modular workbook, and its purpose is to provide you with the necessary skills to perform well in your numerical reasoning test. There is no magic formula for improving your performance in numerical reasoning tests; however, performance is determined by a number of factors aside from basic intelligence. Preparation plays a large part in determining your level of success, and the secret is to practise as much as you can. If the last time you had to work out a percentage increase was a decade ago, it is likely that a quick reminder of the method will help you complete the calculation within the time allowed in the test. This workbook will explain these formulae to remind you how to complete such calculations. Test preparation and good exam technique gives you the confidence to estimate correct answers quickly. Method and practice will help you to calculate correct answers swiftly. This book will help you to prepare for your test by giving you plenty of examples, practice questions and explanations.

What this workbook doesn't do

The content of this workbook is aimed at adult test-takers who want to prepare to take a numerical reasoning test. It is assumed that you don't necessarily want, or have time, to become a mathematician, but you do want to relearn enough maths to help you to do well in the test. This workbook is designed to help you to prepare the required numeracy skills for the standardized aptitude tests currently on the market. This workbook will not explain why mathematical formulae work the way they do. Rather, it explains how to apply the formula in a practical setting, in particular in the application of maths in numeracy tests. It is likely that if you wanted to understand the theory of maths, you would be already immersed in advanced level maths books and wouldn't need this refresher course.

Content of this workbook

Each chapter first explains the concept, works through an example and provides you with practice questions to help consolidate your learning. The emphasis throughout the book is on the practice questions and corresponding explanations.

Chapter 1, 'Review the basics', guides you through the fundamentals and gives you the opportunity to get started with numeracy questions. The worked examples in Chapter 1 demonstrate useful methods to complete calculations quickly.

Chapter 2, 'Fractions and decimals', explains how to work with parts of whole numbers expressed either as fractions or decimals. You will practise the methods to add, subtract, multiply and divide fractions and decimals.

Chapter 3, 'Rates', reminds you of the formulae to work out speed, distance and time when you know two (or more) variables. This chapter also covers the work rate formula.

Chapter 4, 'Percentages', covers the three variables in a typical percentage question. The part, the whole and the percentage are all explained. Percentage increases and decreases are also explained, and you will practise questions involving simple and compound interest.

Chapter 5, 'Ratios and proportions', examines 'part to part' and 'part to whole' type ratios and explains how to use ratios to find actual quantities. An explanation of the use of proportions is included.

Chapter 6, 'Data interpretation', takes all the skills you have practised and combines them to allow you to practise applying your new skills to data presented in graphs, tables and charts.

Chapter 7, 'Glossary of terms and formulae', lists all the terms and formulae you have used throughout the book and serves as a reference guide.

Chapter 8, 'Recommendations for further practice', identifies a number of useful publications and websites for you to use for further practice prior to your test.

In each chapter, an explanation is offered to solve each of the problems. Often there are several different ways to solve a problem. You may find that your preferred method differs from the explanation offered. As you work through more and more examples, you will discover the methods you find easiest to work with, so do try other methods, either to check your answer, or to find out whether there is a quicker way to solve the problem. Remember that speed is a key indicator of success.

How to use this workbook

This workbook is designed to teach the foundation skills relevant to a number of numerical reasoning tests currently on the market. It is progressive and you should work through each of the chapters in order. Concepts explained in the earlier chapters are used and tested again in later chapters. It is good practice for employers and recruiters to send you sample questions prior to the test, in order that you know what to expect on the day. Once you know the broad content of the test, you can use this workbook to practise the areas most relevant to you if time is short. It is recommended that you work thoroughly through Chapter 1 regardless of the type of test you are taking, as Chapter 1 provides you with the skills to build a numerical foundation relevant to subsequent chapters.

Preparation and test technique

Being prepared for the test

If you don't know what kind of questions to expect in your test, there are two things you can do. First, call the tester or recruiter and ask for a sample practice booklet. It is good practice for recruiters to distribute practice tests to prepare you for the type of tests you will be taking. Second, while you are waiting for the practice questions to arrive, practise any sort of maths. Many people find that part of the difficulty they experience with numeracy tests is the lack of familiarity with everyday maths skills. This book aims to help you to overcome that lack of familiarity by providing you with the opportunity to refresh your memory through plenty of practice drills and questions.

Types of question

There are a wide range of questions set in a test. There are also a number of different ways you may be asked to answer the question. Typical answer formats include calculation, multiple choice and data sufficiency answers, where you are given pieces of information and asked whether you have enough information to answer the questions correctly. Chapters 1 to 5 require that you calculate the answers to practice questions. The reason for the adoption of this method is to ensure that you can work out correct answers confidently, without resorting to a 'multiple guess' technique. In Chapter 6, 'Data interpretation', you are given a range of answers from which to choose the correct one. The answer choices include deliberate traps, rather like those the test-writers will set for you. The accompanying explanations will help you to learn about these traps and to help you to avoid them in the test.

Getting to the right answer

When you are asked to calculate an exact answer, calculate it. The question will give you an indication of the level of precision that is expected from you, for example, 'Give the answer to 3 decimal places'. If you are asked for this level of precision, usually you will have a calculator to assist you.

If you are given a range of answers to choose from, do a quick estimate of the correct answer first. Then eliminate all out-of-range answers or the 'outliers'. This technique reduces the likelihood of choosing incorrectly under pressure and gives you a narrower range of answers from which to choose the correct answer. Your estimate may be accurate enough to choose the correct answer without completing any additional calculations. This will save you time and allow you to spend more time on difficult questions. Once you have eliminated some answers from the multiple-choice range, you can substitute in possible correct answers to the question and effectively solve the question using the answer as the starting point.

Translating the language

Part of the difficulty of aptitude tests is in understanding exactly what is being asked of you. Before you set off to answer a question, be absolutely sure that you understand what you are being asked to do. Think carefully whether you have enough information to translate the question into an equation, or whether you need to complete an interim step to provide you with enough information to answer the question. If you are working with graphs and charts, read the labels accompanying the diagrams to make sure you understand whether you are being given percentages or actual values. Read the axis label in case the axes are given in different values. Typically, once you have translated the words, the maths becomes much easier.

Calculators

Many aptitude tests disallow the use of calculators, so you may as well get into the habit of doing mental arithmetic without it. If you are of the GCSE generation, maths without a calculator may seem impossible: after all, calculators are a key tool in maths learning today. However, all the examples and practice questions in this book have been designed so that you can work out the correct answer without the use of the calculator. Think of it as a positive. Think of it this way: if you can't use a calculator in the test, the maths can't be that hard, can it? If you set about the practice questions and drills with your calculator, you will be wasting your test preparation time and practice material. By all means, check your answers afterwards with your calculator, but get into the habit of sitting down to take the test without it.

Test timings

Where timings are applied to the practice questions in this workbook, do try to stick to the time allocated. One of the skills tested in aptitude tests is your ability to work quickly and accurately under pressure. However, sometimes you do not have to finish all the questions in order to do well in a test, but of those you do answer, you must answer those questions correctly. If you find that at first you are taking longer to complete the questions than the allocated time, work out which aspect of the problem is taking the extra time. For example, are you wasting time reworking simple calculations or are you having difficulty in working out exactly what is being asked in the question? Once you have identified the problem, you can go back to the relevant section of this book to find suggestions to help overcome the difficulty.

What else can you do to prepare?

If you find that you draw a complete blank with some of the mental arithmetic questions, particularly the timed questions, find other ways to exercise your grey matter even when you are not in study mode. Add up the bill in your head as you are grocery shopping. Work out the value of discounts offered in junk e-mail. Work out whether the deal on your current credit card is better or worse than the last one. Work out dates backwards. For example, if today is Saturday 3 August, what was the date last Tuesday? If my birthday is 6 April 1972 and today is Monday, on what day was I born? On the bus on the way to work, work out roughly how many words there must be on the front page of your neighbour's newspaper, based on your estimate of the number of words per line and the number of lines per page. Play sudoko, all levels. In other words, become proficient at estimating everyday calculations. Think proactively about numbers and mental arithmetic. It will pay off enormously in the test.

Getting started

This introduction has explained the purpose of this workbook and has recommended a method to use it. Now roll up your sleeves and go straight on to Chapter 1.

1

Review the basics

Chapter topics

- Terms used in this chapter
- Multiplication tables
- Dividing and multiplying numbers
- Prime numbers
- Multiples
- Working with large numbers
- Working with signed numbers
- Averages
- Answers to Chapter 1

Terms used in this chapter

Arithmetic mean: The amount obtained by adding two or more numbers and dividing by the number of terms.

Average: See Mode, Median and Arithmetic mean.

Dividend: The number to be divided.

Divisor: The number by which another is divided.

Factor: The positive integers by which an integer is evenly divisible.

Find the product of ... : Multiply two or more numbers together.
Integer: A whole number without decimal or fraction parts.
Lowest common multiple: The least quantity that is a multiple of two or more given values.
Mean: See Arithmetic mean.
Median: The middle number in a range of numbers when the set is arranged in ascending or descending order.
Mode: The most popular value in a set of numbers.
Multiple: A number that divides into another without a remainder.
Prime factor: The factors of an integer that are prime numbers.
Prime number: A number divisible only by itself and 1.

Test-writers assume that you remember the fundamentals you learnt in school and that you can apply that knowledge and understanding to the problems in the tests. The purpose of this chapter is to remind you of the basics and to provide you with the opportunity to practise them before your test. The skills you will learn in this chapter are the fundamentals you can apply to solving many of the problems in an aptitude test, so it is worth learning the basics thoroughly. You must be able to do simple calculations very quickly, without expending any unnecessary brainpower – keep this in reserve for the tricky questions later on. This chapter reviews the basics and includes a number of practice drills to ease you back into numerical shape. Remember, no calculators ...

Multiplication tables

'Rote learning' as a teaching method has fallen out of favour in recent years. There are good reasons for this in some academic areas but it doesn't apply to multiplication tables. You learnt the times-tables when you first went to school, but can you recite the tables as quickly now? Recite them to yourself quickly, over and over again, when you're out for a run, when you're washing up, when you're cleaning your teeth, when you're stirring your baked beans -- any time when you have a spare 10 seconds thinking time. Six times, seven times and eight times are the easiest to forget, so drill these more often than the twos and fives. Make sure that you can respond to any multiplication question without pausing even for half a second. If you know the multiplication tables inside out, you will save yourself valuable seconds in your test and avoid needless mistakes in your calculations.

Multiplication tables: practice drill 1

Practise these drills and aim to complete each set within 15 seconds. (Remember, the answers are at the end of the chapter.)

	Drill 1	Drill 2	Drill 3	Drill 4	Drill 5
1	3 × 7 =	8 × 3 =	9 × 3 =	7 × 5 =	7 × 15 =
2	6 × 5 =	11 × 6 =	9 × 2 =	6 × 3 =	11 × 11 =
3	8 × 9 =	13 × 2 =	7 × 7 =	4 × 7 =	4 × 12 =
4	3 × 3 =	11 × 13 =	12 × 7 =	3 × 4 =	8 × 10 =
5	9 × 12 =	13 × 9 =	6 × 8 =	8 × 15 =	13 × 4 =
6	2 × 4 =	6 × 14 =	6 × 7 =	8 × 8 =	11 × 2 =
7	8 × 5 =	3 × 15 =	13 × 5 =	2 × 6 =	7 × 12 =
8	13 × 3 =	9 × 8 =	13 × 4 =	7 × 6 =	9 × 15 =
9	6 × 7 =	4 × 5 =	8 × 8 =	4 × 12 =	9 × 9 =
10	2 × 7 =	6 × 3 =	7 × 13 =	5 × 14 =	3 × 8 =
11	7 × 12 =	12 × 4 =	6 × 15 =	3 × 11 =	3 × 3 =
12	2 × 12 =	11 × 7 =	9 × 8 =	9 × 6 =	3 × 8 =

Multiplication tables: practice drill 2

	Drill 1	Drill 2	Drill 3	Drill 4	Drill 5
1	8 × 9 =	4 × 5 =	3 × 6 =	6 × 11 =	11 × 11 =
2	13 × 13 =	9 × 3 =	9 × 5 =	5 × 3 =	3 × 7 =
3	11 × 14 =	13 × 4 =	7 × 3 =	5 × 6 =	6 × 8 =
4	9 × 6 =	6 × 8 =	5 × 9 =	8 × 5 =	12 × 4 =
5	11 × 5 =	7 × 7 =	11 × 15 =	14 × 14 =	5 × 5 =
6	4 × 3 =	11 × 10 =	6 × 7 =	12 × 3 =	8 × 14 =
7	7 × 8 =	8 × 9 =	3 × 11 =	14 × 4 =	9 × 7 =
8	9 × 6 =	5 × 13 =	8 × 4 =	13 × 3 =	12 × 15 =
9	12 × 5 =	3 × 14 =	9 × 6 =	7 × 6 =	3 × 3 =
10	13 × 14 =	2 × 12 =	12 × 4 =	8 × 7 =	4 × 2 =
11	4 × 9 =	4 × 6 =	8 × 2 =	3 × 8 =	14 × 8 =
12	2 × 5 =	9 × 5 =	3 × 13 =	11 × 12 =	6 × 10 =

Dividing and multiplying numbers

Long multiplication

Rapid multiplication of multiple numbers is easy if you know the multiplication tables inside-out and back-to-front. In a long multiplication calculation, you break the problem down into a number of simple calculations by dividing the multiplier up into units of tens, hundreds, thousands and so on. In Chapter 2 you will work through practice drills involving division and multiplication of decimals.

Worked example

Q. What is the result of 2,348 × 237?

To multiply a number by 237, break the problem down into a number of simpler calculations. Divide the multiplier up into units of hundreds, tens and units.

For example, to multiply by 237 you multiply by:

7 (units)
3 (tens)
2 (hundreds)
(It doesn't matter in which order you complete the calculation.)

2,348		
237	× =	Multiplier
16,436	=	2,348 × 7
70,440	=	2,348 × 30
469,600	=	2,348 × 200
556,476	=	16,436 + 70,440 + 469,600

Long multiplication: practice drill 1

No calculators! This exercise is intended to help you to speed up your mental arithmetic. Set a stopwatch and aim to complete this drill in five minutes.

Q1	12	×	24
Q2	13	×	23
Q3	11	×	23
Q4	19	×	19
Q5	26	×	24
Q6	213	×	43
Q7	342	×	45
Q8	438	×	23
Q9	539	×	125
Q10	5,478	×	762

Long multiplication: practice drill 2

Set a stopwatch and aim to complete this practice drill in five minutes.

Q1	9	×	18
Q2	11	×	19
Q3	12	×	21
Q4	19	×	23
Q5	26	×	19
Q6	211	×	17
Q7	317	×	13
Q8	416	×	11
Q9	624	×	97
Q10	725	×	101

Long division

Long division calculations, like long multiplication calculations, can be completed quickly and easily without a calculator if you know the multiplication tables well. There are four steps in a long division calculation, and as long as you follow these in order, you will arrive at the right answer.

Worked example

Q: Divide 156 by 12

This may seem obvious, but recognize which number you are dividing *into*. This is called the *dividend*. In this case you are dividing the dividend (156) by the *divisor* (12). Be clear about which is the divisor and which is the dividend – this will become very important when you divide very large or very small numbers.

There are four steps in a long division question.

Step 1	Divide (**D**)
Step 2	Multiply (**M**)
Step 3	Subtract (**S**)
Step 4	Bring down (**B**)

You can remember this as D-M-S-B with any mnemonic that helps you to remember the order. Do-Mind-Slippery-Bananas. Dirty-Muddy-Salty-Bicycles. Follow the steps in order and repeat until you have worked through the whole calculation.

Step 1: Divide

Work from the left to the right of the whole number. 12 divides into 15 once, so write '1' on top of the division bar.

$$\begin{array}{r} 1 \\ 12\,\overline{)156} \end{array}$$

Step 2: Multiply

Multiply the result of step 1 (1) by the divisor (12):

1 × 12 = 12. Write the number 12 directly under the dividend (156).

$$\begin{array}{r} 1 \\ 12\,\overline{)156} \\ \underline{12} \end{array}$$

Step 3: Subtract

Subtract 12 from 15 and write the result directly under the result of Step 2.

$$\begin{array}{r} 1 \\ 12\,\overline{)156} \\ \underline{-12} \\ 3 \end{array}$$

Step 4: Bring down

Bring down the next digit of the dividend (6).

```
      1
12 ) 156
    –12(
    ----
     36
```

Return to Step 1 and start the four-step process again.

Step 1: Divide 12 into 36 and write the result (3) on top of the long division sign.

```
     13
12 ) 156
   – 12(
    ----
     36
```

Step 2: Multiply the result of Step 1 (3) by the divisor (12): 3 × 12 = 36. Write the number 36 directly below the new dividend (36).

```
     13
12 ) 156
   – 12(
    ----
     36
     36
    ----
```

Step 3: Subtract 36 from 36.

```
      13
12 ) 156
    –12(
    ----
      36
    – 36
    ----
       0
```

Step 4: There aren't any more digits to Bring down, so the calculation is complete.

156 ÷ 12 = 13

You will learn about long division with remainders and decimals in Chapter 2.

Long division: practice drill 1

Set a stopwatch and aim to complete these calculations in four minutes. You may check your answers with a calculator only once you have finished all the questions in the drill.

Q1	99	÷	11
Q2	91	÷	13
Q3	117	÷	13
Q4	182	÷	14
Q5	696	÷	58
Q6	3,024	÷	27
Q7	2,890	÷	34
Q8	636	÷	53
Q9	1,456	÷	13
Q10	2,496	÷	78

Long division: practice drill 2

Set a stopwatch and aim to complete the following practice drill within five minutes.

Q1	1,288	÷	56
Q2	1,035	÷	45
Q3	6,328	÷	56
Q4	5,625	÷	125
Q5	2,142	÷	17
Q6	7,952	÷	142
Q7	10,626	÷	231
Q8	11,908	÷	458
Q9	81,685	÷	961
Q10	3,591	÷	27

Prime numbers

An integer greater than 1 is a prime number if its only positive divisors are itself and 1. All prime numbers apart from 2 are odd numbers. Even numbers are divisible by 2 and cannot be prime by definition. 1 is not a prime number, because it is divisible by one number only, itself. The following is a list of all the prime numbers below 100. It's worth becoming familiar with these numbers so that when you come across them in your test, you don't waste time trying to find other numbers to divide into them!

0–10	2 3 5 7
11–20	11 13 17 19
21–30	23 29
31–40	31 37
41–50	41 43 47
51–60	53 59

61–70	61 67
71–80	71 73 79
81–90	83 89
91–100	97

Prime numbers: practice drill

Refer to the table above to assist you with the following drill:

Q1 What is the product of the first four prime numbers?
Q2 What is the sum of the prime numbers between 40 and 50 minus the eleventh prime number?
Q3 How many prime numbers are there?
Q4 How many prime numbers are there between 1 and 100?
Q5 What is the only even prime number?
Q6 Between 1 and 100, there are five prime numbers ending in 1. What are they?
Q7 What is the result of the product of the first three prime numbers minus the sum of the second three prime numbers?
Q8 How many prime numbers are there between 60 and 80?
Q9 How many prime numbers are there between 90 and 100?
Q10 What is the sum of the second 12 prime numbers minus the sum of the first 12 prime numbers?

Multiples

A *multiple* is a number that divides by another without a remainder. For example, 54 is a multiple of 9 and 72 is a multiple of 8.

Tips to find multiples

An integer is divisible by:

2, if the last digit is 0 or is an even number
3, if the sum of its digits are a multiple of 3
4, if the last two digits are a multiple of 4
5, if the last digit is 0 or 5
6, if it is divisible by 2 and 3
9, if its digits sum to a multiple of 9

There is no consistent rule to find multiples of 7 or 8.

Worked example

Is 2,648 divisible by 2? Yes, because 8 is divisible by an even number.
Is 91,542 divisible by 3? Yes, because 9+1+5+4+2 = 21 and 21 is a multiple of 3.
Is 216 divisible by 4? Yes, because 16 is a multiple of 4.
Is 36,545 divisible by 5? Yes, because the last digit is 5.
Is 9,918 divisible by 6? Yes, because the last digit, 8, is divisible by an even number and the sum of all the digits, 27, is a multiple of 3.

Multiples: practice drill

Set a stopwatch and aim to complete the following 10-question drill in five minutes.

The following numbers are multiples of which of the following integers: 2, 3, 4, 5, 6, 9?

	Drill 1	Drill 2	Drill 3	Drill 4
1	36	2,654	642	542
2	218	23	8,613	9,768
3	5,244	96	989,136	8,752
4	760	524	652	92
5	7,735	97	1,722	762
6	29	152	13	276
7	240,702	17,625	675	136
8	81,070	7,512	124	19
9	60,472	64	86	9005
10	161,174	128	93	65

Lowest common multiple

The *lowest common multiple* is the least quantity that is a multiple of two or more given values. To find a multiple of two integers, you can simply multiply them together, but this will not necessarily give you the lowest common multiple of both integers. To find the lowest common multiple, you will work with the prime numbers. This is a concept you will find useful when working with fractions. There are three steps to find the lowest common multiple of two or more numbers:

Step 1: Express each of the integers as the product of its prime factors.
Step 2: Line up common prime factors.
Step 3: Find the product of the distinct prime factors.

Worked example

What is the lowest common multiple of 6 and 9?

Step 1: Express each of the integers as the product of its prime factors

To find the prime factors of an integer, divide that number by the prime numbers, starting with 2. The product of the prime factors of an integer is called the prime factorization.

Divide 6 by 2:

$$\begin{array}{r} 3 \\ 2\overline{)6} \end{array}$$

Now divide the remainder, 3, by the next prime factor after 2:

$$\begin{array}{r} 1 \\ 3\overline{)3} \end{array}$$

So the prime factors of 6 are 2 and 3. (Remember that 1 is not a prime number.) The product of the prime factors of an integer is called the prime factorization, so the prime factorization of 6 = 2 × 3.

Now follow the same process to work out the prime factorization of 9 by the same process. Divide 9 by the first prime number that divides without a remainder:

$$\begin{array}{r} 3 \\ 3\overline{)9} \end{array}$$

Now divide the result by the first prime number that divides without a remainder.

$$\begin{array}{r} 1 \\ 3\overline{)3} \end{array}$$

The prime factorization of 9 = 3 × 3.

Step 2: Line up common prime factors

Line up the prime factors of each of the given integers below each other:

$$6 = 2 \times 3$$
$$9 = 3 \times 3$$

Notice that 6 and 9 have a common prime factor (3).

Step 3: Find the product of the prime factors

Multiply all the prime factors together. When you see a common prime factor, count this only once.

$$6 = 2 \times \textcircled{3}$$
$$9 = \textcircled{3} \times 3$$

Prime factorization =

$$2 \times 3 \times 3 = 18$$

The lowest common multiple of 6 and 9 = 18.

Lowest common multiple: practice drill 1

Set a stopwatch and aim to complete the following drill in four minutes. Find the lowest common multiple of the following sets of numbers:

Q1	8 and 6
Q2	12 and 9
Q3	3 and 5
Q4	12 and 15
Q5	8 and 14
Q6	9 and 18
Q7	4 and 7

Q8	13 and 7
Q9	12 and 26
Q10	7 and 15

Lowest common multiple: practice drill 2

Set a stopwatch and aim to complete the following drill in four minutes. Find the lowest common multiple of the following sets of numbers:

Q1	2 and 5
Q2	4 and 5
Q3	5 and 9
Q4	6 and 5
Q5	6 and 7
Q6	2 and 5 and 6
Q7	3 and 6 and 7
Q8	3 and 7 and 8
Q9	3 and 6 and 11
Q10	4 and 6 and 7

Working with large numbers

Test-writers sometimes set questions that ask you to perform an operation on very large or very small numbers. This is a cruel test trap, as it is easy to be confused by a large number of decimal places or zeros. Operations on small and large numbers are dealt with in Chapter 2. This section reminds you of some commonly used terms and their equivalents.

Millions, billions and trillions

The meaning of notations such as millions, billions and trillions is ambiguous. The terms vary and the UK definition of these terms is different from the US definition. If you are taking a test developed in the United States (such as the GMAT or GRE), make sure you know the difference.

US definition

Million: 1,000,000 or 'a thousand thousand' (same as UK definition)
Billion: 1,000,000,000 or 'a thousand million'
Trillion: 1,000,000,000,000 or 'a thousand billion'

UK definition

Million: 1,000,000 or 'a thousand thousand' (same as US definition)
Billion: 1,000,000,000,000 or 'a million million'
Trillion: 1,000,000,000,000,000,000 or 'a million million million'

The US definitions are more commonly used now. If you are in doubt and do not have the means to clarify which notation is being used, assume the US definition.

Multiplying large numbers

To multiply large numbers containing enough zeros to make you go cross-eyed, follow these three steps:

Step 1: Multiply the digits greater than 0 together.
Step 2: Count up the number of zeros in each number.
Step 3: Add that number of zeros to the result of Step 1.

Worked example

What is the result of 2,000,000 × 2,000?

Step 1: Multiply the digits greater than 0 together
2 × 2 = 4

Step 2: Count up the number of zeros in each number
2,000,000 × 2,000 = nine zeros

Step 3: Add that number of zeros to the right of the result of Step 1
Result of Step 1 = 4
nine zeros = 000,000,000
Answer = 4,000,000,000 (4 billion, US definition)

Multiplying large numbers: practice drill

Use the US definition of billion and trillion to complete this practice drill. Set a stopwatch and aim to complete the following drill in three minutes.

Q1 9 hundred × 2 thousand
Q2 2 million × 3 million
Q3 3 billion × 1 million
Q4 12 thousand × 4 million
Q5 24 million × 2 billion
Q6 18 thousand × 2 million
Q7 2 thousand × 13 million
Q8 3 hundred thousand × 22 thousand
Q9 28 million × 12 thousand
Q10 14 billion × 6 thousand

Dividing large numbers

Divide large numbers in exactly the same way as you would smaller numbers, but cancel out equivalent zeros before you start.

Worked example

4,000,000 ÷ 2,000

Cancel out equivalent zeros:

4,000,~~000~~ ÷ 2,~~000~~

Now you are left with an easier calculation:

4,000 ÷ 2

Answer = 2,000

Dividing large numbers: practice drill

Set a stopwatch and aim to complete the following drill in four minutes.

Q1	8,000	÷	20
Q2	2,700	÷	90
Q3	240,000	÷	600
Q4	6,720,000	÷	5,600
Q5	475,000	÷	1,900
Q6	19,500,000	÷	15,000
Q7	23,800,000	÷	140
Q8	149,500,000	÷	65,000
Q9	9,890,000	÷	2,300
Q10	15,540,000	÷	42,000

Working with signed numbers

Multiplication of signed numbers

There are a few simple rules to remember when multiplying signed numbers.

Positive × positive = positive	P × P = P
Negative × negative = positive	N × N = P
Negative × positive = negative	N × P = N
Positive × negative = negative	P × N = N

Tip: note that it doesn't matter which sign is presented first in a multiplication calculation.

The product of an odd number of negatives = negative
N × N × N = N

The product of an even number of negatives = positive
N × N × N × N = P

Worked example

P × P = P	2 × 2 = 4
N × N = P	–2 × –2 = 4
N × P = N	–2 × 2 = –4
P × N = N	2 × –2 = –4
N × N × N = N	–2 × –2 × –2 = –8
N × N × N × N = P	–2 × –2 × –2 × –2 = 16

Division of signed numbers

Positive ÷ positive = positive	P ÷ P = P
Negative ÷ negative = positive	N ÷ N = P
Negative ÷ positive = negative	N ÷ P = N
Positive ÷ negative = negative	P ÷ N = N

Worked example

P ÷ P = P	2 ÷ 2 = 1
N ÷ N = P	–2 ÷ –2 = 1
P ÷ N = N	2 ÷ –2 = –1
N ÷ P = N	–2 ÷ 2 = –1

Multiplication and division of signed numbers: practice drill I

Set a stopwatch and aim to complete each drill within five minutes.

Q1	12	×	12
Q2	– 12	×	– 12
Q3	14	×	– 3
Q4	27	×	– 13
Q5	– 19	×	19
Q6	– 189	÷	21
Q7	– 84	÷	– 6
Q8	1,440	÷	– 32
Q9	– 221	÷	– 17
Q10	414	÷	– 23

Multiplication and division of signed numbers: practice drill 2

Q1	16	×	13
Q2	27	×	– 29
Q3	– 131	×	21
Q4	– 52	×	– 136
Q5	272	×	– 13
Q6	–112	÷	2
Q7	–72	÷	– 24
Q8	540	÷	12
Q9	4,275	÷	– 19
Q10	– 3,638	÷	– 214

Averages

One way to compare sets of numbers presented in tables, graphs or charts is by working out the average. This is a technique used in statistical analysis to analyse data and to draw conclusions about the content of the data set. The three types of averages are the *arithmetic mean*, the *mode* and the *median*.

Arithmetic mean

The arithmetic mean (also known simply as the average) is a term you are probably familiar with. To find the mean, simply add up all the numbers in the set and divide by the number of terms.

$$\textit{Arithmetic mean} = \frac{\textit{Sum of values}}{\textit{Number of values}}$$

Worked example

In her aptitude test, Emma scores 77, 81 and 82 in each section. What is her average (arithmetic mean) score?

$$\textit{Arithmetic mean} = \frac{(77 + 81 + 82) = 240}{3}$$

Arithmetic mean = 80

Worked example

What is the arithmetic mean of the following set of numbers: 0, 6, 12, 18?

$$\textit{Arithmetic mean} = \frac{(0 + 6 + 12 + 18) = 36}{4}$$

The arithmetic mean of the set is 36 ÷ 4 = 9.

Tip: remember to include the zero as a value in your sum of the number of values.

Worked example

What is the value of q if the arithmetic mean of 3, 6, 9 and q = 5.5?

Step 1: Rearrange the formula to help you to find the sum of values

Sum of values = arithmetic mean × number of values

Step 2: Now plug in the numbers

Sum of values = arithmetic mean × number of values

Sum of values = 5.5 × 4 = 22

(Don't forget to count the fourth value *q* in the number of values.)

Sum of values = 22

Step 3: Subtract the sum of known values from the sum of values

Sum of values – sum of known values = q

22 – 18 (3 + 6 + 9) = q

Answer: q = 4

The mode

The mode is the number (or numbers) that appear(s) the most frequently in a set of numbers. There may be more than one mode in a given set of numbers.

Worked example

What is the mode in the following set of numbers?

21, 22, 23, 22, 22, 25, 25, 22, 21

21 appears twice.
22 appears four times.
23 appears once.
25 appears twice.
So the mode is 22 as it appears most frequently.

Worked example

What is the mode in the following set of numbers?

–0.01, 0.01, –0.01, –0.1, 0.01, –0.01, 0.1, 0.01, –0.1

0.1 appears once.
0.01 appears three times.
–0.1 appears twice.
–0.01 appears three times.
So the mode numbers are 0.01 and –0.01.

The median

The median is the value of the middle number in a set of numbers, when the numbers are put in ascending or descending order.

Worked example

What is the median in the following set of numbers?

82, 21, 34, 23, 12, 46, 65, 45, 37

First order the numbers in ascending (or descending) order.

12, 21, 23, 34, 37, 45, 46, 65, 82

As there are nine numbers in the set, the fifth number in the series is the median.

1	*2*	*3*	*4*	*5*	*6*	*7*	*8*	*9*
12	21	23	34	37	45	46	65	82

37 is the median value.

Worked example

What is the median in the following set of numbers?

0, 2, 6, 10, 4, 8, 0, 1

First put the numbers of the set in order (either ascending or descending).

0, 0, 1, 2, 4, 6, 8, 10

This time there is an even number of values in the set.

1	*2*	*3*	*4*	*5*	*6*	*7*	*8*
0	0	1	2	4	6	8	10

Draw a line in the middle of the set:

1	*2*	*3*	*4*	*5*	*6*	*7*	*8*
0	0	1	2	4	6	8	10

The median of the series is the average of the two numbers on either side of the dividing line. Therefore, the median number in the series is the arithmetic mean of 2 and 4:
$(2 + 4) \div 2 = 3$.

Averages: practice drill

Set a stopwatch and aim to complete the following drill in 60 seconds.

What is the arithmetic mean of the following sets of numbers?

Q1 17, 18, 21, 23, 21
Q2 0.1, 0.2, 0.2, 0.3, 0.3, 0.1
Q3 0, 2, 4, 6
Q4 –2, –4, –6, –10
Q5 –0.2, 0.1, 0.3, 0.6, 0.7, 0.9

What is the median of the following sets of numbers?

Q6 8, 1, 6, 2, 4
Q7 –6, –2.5, 0, –1, –3
Q8 82, 73, 72, 72, 71
Q9 6, 0, 3, 9, 15, 12
Q10 36, 32, 37, 41, 39, 39

What is the mode in the following sets of numbers?

Q11 21, 22, 23, 23, 22, 23
Q12 –2, –7, –2, –7, –7, –4, –4, –2, –3, –2
Q13 0, 1, 1, 1, 0, 0, 0, 1, 1
Q14 $\frac{1}{2}, \frac{1}{4}, \frac{3}{4}, \frac{3}{4}, \frac{1}{4}, \frac{1}{2}, \frac{1}{4}$
Q15 –0.1, 0.001, 0.01, –0.1, 0.001, –0.1, 0.001

Answers to Chapter 1

Multiplication tables: practice drill 1

	Drill 1	Drill 2	Drill 3	Drill 4	Drill 5
1	21	24	27	35	105
2	30	66	18	18	121
3	72	26	49	28	48
4	9	143	84	12	80
5	108	117	48	120	52
6	8	84	42	64	22
7	40	45	65	12	84
8	39	72	52	42	135
9	42	20	64	48	81
10	14	18	91	70	24
11	84	48	90	33	9
12	24	77	72	54	24

Multiplication tables: practice drill 2

	Drill 1	Drill 2	Drill 3	Drill 4	Drill 5
1	72	20	18	66	121
2	169	27	45	15	21
3	154	52	21	30	48
4	54	48	45	40	48
5	55	49	165	196	25
6	12	110	42	36	112
7	56	72	33	56	63
8	54	65	32	39	180
9	60	42	54	42	9
10	182	24	48	56	8
11	36	24	16	24	112
12	10	45	39	132	60

Long multiplication: practice drill 1

Q1	288
Q2	299
Q3	253
Q4	361
Q5	624
Q6	9,159
Q7	15,390
Q8	10,074
Q9	67,375
Q10	4,174,236

Long multiplication practice: drill 2

Q1	162
Q2	209
Q3	252
Q4	437
Q5	494
Q6	3,587
Q7	4,121
Q8	4,576
Q9	60,528
Q10	73,225

Long division practice: drill 1

Q1	9
Q2	7
Q3	9
Q4	13
Q5	12

Q6	112
Q7	85
Q8	12
Q9	112
Q10	32

Long division practice: drill 2

Q1	23
Q2	23
Q3	113
Q4	45
Q5	126
Q6	56
Q7	46
Q8	26
Q9	85
Q10	133

Prime numbers: practice drill

Q1	210
Q2	100
Q3	An infinite number
Q4	25
Q5	2
Q6	11, 31, 41, 61, 71
Q7	–1
Q8	5
Q9	1
Q10	569

Multiples: practice drill

	Drill 1	Drill 2	Drill 3	Drill 4
1	2,3,4,6,9	2	2,3,6	2
2	2	Prime	3,9	2,3,4,6 [8]
3	2,3,4,6	2,3,4,6	2,3,4,6,9 [8]	2,4 [8]
4	2,4,5	2,4	2,4	2,4
5	5 [7]	Prime	2,3,6 [7]	2,3,6
6	Prime	2,4 [8]	Prime	2,3,4,6
7	2,3,6	3,5	3,5,9	2,4 [8]
8	2,5	2,3,4,6 [8]	2,4	Prime
9	2,4 [8]	2,4 [8]	2	5
10	2	2,4 [8]	3	5

Where the integer is also a multiple of 7 or 8, this is indicated in the answer table with [7] or [8].

Lowest common multiple: practice drill 1

Q1 8 and 6
Answer = 24

8	=	2	×	2	×	(2)		
6	=					(2)	×	3

Prime factorization = 2 × 2 × 2 × 3

Q2 12 and 9
Answer = 36

12	=	2	×	2	×	(3)		
9	=					(3)	×	3

Prime factorization = 2 × 2 × 3 × 3

Q3 3 and 5
Answer = 15

3	=	(1)	×	3		
5	=	(1)	×	5		

Prime factorization = (1 ×) 3 × 5

Q4 12 and 15
Answer = 60

12	=	2	×	2	×	(3)		
15	=					(3)	×	5

Prime factorization = 2 × 2 × 3 × 5

Q5 8 and 14
Answer = 56

8	=	2	×	2	×	(2)		
14	=					(2)	×	7

Prime factorization = 2 × 2 × 2 × 7

Q6 9 and 18
Answer = 18

9	=			(3)	×	(3)
18	=	2	×	(3)	×	(3)

Prime factorization = 2 × 3 × 3

Q7 4 and 7
Answer = 28

4	=	2	×	2
7	=	1	×	7

Prime factorization = (1 ×) 2 × 2 × 7

Q8 13 and 7
Answer = 91

13	=	1	×	13
7	=	1	×	7

Prime factorization = (1 ×) 13 × 7

Q9 12 and 26
Answer = 156

12	=	2	×	2	×	3
26	=			2	×	13

Prime factorization = 2 × 2 × 3 × 13

Q10 7 and 15
Answer = 105

7	=	1	×	7
15	=	3	×	5

Prime factorization = (1 ×) 3 × 5 × 7

Lowest common multiple: practice drill 2

Q1	10
Q2	20
Q3	45
Q4	30
Q5	42
Q6	30
Q7	42
Q8	168
Q9	66
Q10	84

Multiplying large numbers: practice drill

Q1	1 million, 8 hundred thousand	(1,800,000)
Q2	6 trillion	(6,000,000,000,000)
Q3	3 thousand trillion	(3,000,000,000,000,000)
Q4	48 billion	(48,000,000,000)
Q5	48 thousand trillion	(48,000,000,000,000,000)
Q6	36 billion	(36,000,000,000)
Q7	26 billion	(26,000,000,000)
Q8	6 billion, 6 hundred million	(6,600,000,000)
Q9	336 billion	(336,000,000,000)
Q10	84 trillion	(84,000,000,000,000)

Dividing large numbers: practice drill

Q1	400
Q2	30
Q3	400
Q4	1,200
Q5	250

Q6	1,300
Q7	170,000
Q8	2,300
Q9	4,300
Q10	370

Multiplication and division of signed numbers: practice drill 1

Q1	144
Q2	144
Q3	–42
Q4	–351
Q5	–361
Q6	–9
Q7	14
Q8	–45
Q9	13
Q10	–18

Multiplication and division of signed numbers: practice drill 2

Q1	208
Q2	–783
Q3	–2,751
Q4	7,072
Q5	–3,536
Q6	–56
Q7	3
Q8	45
Q9	–225
Q10	17

Averages: practice drill

Arithmetic mean:

Q1	20
Q2	0.2
Q3	3
Q4	–5.5
Q5	0.4

Median:

Q6	4
Q7	–2.5
Q8	72
Q9	7.5
Q10	38

Mode:

Q11	23
Q12	–2
Q13	1
Q14	$\frac{1}{4}$
Q15	–0.1 and 0.001

2

Fractions and decimals

Chapter topics

- Terms used in this chapter
- What a fraction is
- Working with fractions
- Fraction operations
- Decimal operations
- Answers to Chapter 2

Terms used in this chapter

Denominator: The number below the line in a vulgar fraction.
Dividend: The number to be divided.
Divisor: The number by which another is divided.
Equivalent fractions: Fractions with equivalent denominators and numerators.
Fraction: A part of a whole number.
Fraction bar: The line that separates the numerator and denominator in a vulgar fraction.

Improper fraction: A fraction in which the numerator is greater than or equal to the denominator.

Lowest common denominator: The lowest common multiple of the denominators of several fractions.

Lowest common multiple: The least quantity that is a multiple of two or more given values.

Mixed fractions: A number consisting of an integer and a fraction.

Numerator: The number above the line in a vulgar fraction.

Prime factorization: The expression of a number as the product of its prime numbers.

Proper fraction: A fraction less than one, with the numerator less than the denominator.

Vulgar fraction: A fraction expressed by numerator and denominator, rather than decimally.

In Chapter 1, you practised operations with whole numbers. In this chapter you will practise number operations on parts of numbers. The same principles apply to decimals and fractions as to whole numbers. Additionally, there are a few extra tricks you can learn to complete these puzzles quickly and accurately.

What a fraction is

Proper and improper fractions

A fraction is a part of a whole number, or a value expressed as one number divided by another. For example:

$\frac{1}{2}$ Part = 1 and whole = 2.

$\frac{2}{3}$ Part = 2 and whole = 3.

$\frac{7}{5}$ Part = 7 and whole = 5.

To write a fraction, put the part over the whole and separate with a fraction bar. The number above the fraction bar is called the *numerator* and the number below the fraction bar is called the *denominator*. There are two types of fractions. A *proper fraction* is a fraction less than one, with the numerator less than the denominator. An *improper fraction* is a fraction in which the numerator is greater than or equal to the denominator. For example:

$\frac{1}{4}$	Numerator = 1 and denominator = 4	Proper fraction
$\frac{2}{5}$	Numerator = 2 and denominator = 5	Proper fraction
$\frac{9}{4}$	Numerator = 9 and denominator = 4	Improper fraction

These terms will become useful when you start to work with fractions.

Working with fractions

Finding the lowest common multiple

In Chapter 1, you practised working out the lowest common multiple. Here's a quick refresher of this method – this is a great technique to have mastered as it will save you time when you're working out fractions under time pressure. Review Chapter 1 now if you need a reminder of the method to find the lowest common multiple.

Finding the lowest common denominator

To find the lowest common denominator of two or more fractions, find the lowest common multiple of all the denominators.

Worked example

What is the lowest common denominator of the following fractions?

$\frac{2}{3} \quad \frac{5}{6} \quad \frac{7}{9}$

Find the lowest common multiple by finding the prime factorization of each denominator:

$3 = 1 \times 3$
$6 = 2 \times 3$
$9 = 3 \times 3$

The lowest common multiple = $(1 \times)\ 2 \times 3 \times 3 = 18$.

Comparing positive fractions

Often you can estimate which of two fractions is the larger. However, if the fractions are too close to estimate accurately, there is a method you can use to work out the relative sizes of fractions. You can also use this technique to find out whether two fractions are equivalent. This is a technique you will find useful to compare ratios and proportions. To compare positive fractions, simply cross-multiply the numerators and denominators.

Worked example

Are the following fractions equivalent?

$$\frac{4}{9} = \frac{12}{27} \quad ?$$

Cross-multiply denominators and numerators:

$$(108)\ \frac{4}{9} \quad \times \quad \frac{12}{27}\ (108)$$

Follow the arrows and write the results beside the fraction.

$4 \times 27 = 108$ and $12 \times 9 = 108$, so the fractions are equivalent.

Worked example

Which is the larger fraction?

$$\frac{5}{9} \ \textit{or}\ \frac{7}{11}$$

$$(55)\ \frac{5}{9} \quad \times \quad \frac{7}{11}\ (63)$$

$5 \times 11 = 55$ and $9 \times 7 = 63$. Follow the arrows and write your answers on either side of the fractions. The fraction on the right side of the question is the larger fraction, because $63 > 55$.

Comparing positive fractions: practice drill

Decide which is the larger fraction.

Q1	$\frac{2}{3}$ *or* $\frac{13}{19}$	Q2	$\frac{3}{4}$ *or* $\frac{13}{16}$
Q3	$\frac{7}{9}$ *or* $\frac{5}{7}$	Q4	$\frac{3}{9}$ *or* $\frac{3}{7}$
Q5	$\frac{4}{17}$ *or* $\frac{2}{9}$	Q6	$\frac{8}{15}$ *or* $\frac{5}{9}$
Q7	$\frac{28}{51}$ *or* $\frac{8}{14}$	Q8	$\frac{2}{50}$ *or* $\frac{7}{175}$
Q9	$\frac{2}{7}$ *or* $\frac{12}{42}$	Q10	$\frac{100}{4}$ *or* $\frac{4}{1}$

Reducing a fraction to its lowest terms

In numerical reasoning tests, you will typically see fractions in the lowest terms. There are two methods to reduce a fraction to its lowest terms.

Method 1

Continue to divide the numerator and denominator by common factors until neither the numerator nor denominator can be factored any further.

Worked example

Reduce $\frac{36}{90}$ to its lowest terms

36 and 90 have a common factor of 18, so divide the numerator and denominator by 18.

$\frac{\not{36}}{\not{90}}$ $\frac{2}{5}$

Method 2

Use the prime factors to reduce a fraction to its lowest terms. This is a useful technique to use when the numbers are large.

Worked example

Reduce $\frac{2310}{3510}$ to its lowest terms.

Factor out the numerator and the denominator:

$$\frac{2310 = 2 \times 3 \times 5 \times 7 \times 11}{3510 = 2 \times 3 \times 5 \times 9 \times 13}$$

Now cancel out common factors:

$$\frac{2310 = \not{2} \times \not{3} \times \not{5} \times 7 \times 11 = 77}{3510 = \not{2} \times \not{3} \times \not{5} \times 9 \times 13 = 117}$$

$$\frac{2310}{3510} = \frac{77}{117}$$

Reducing fractions: practice drill

Q1 $\frac{12}{14} =$

Q2 $\frac{15}{35} =$

Q3 $\frac{33}{39} =$

Q4 $\frac{30}{126} =$

Q5 $\frac{36}{42} =$

Q6 $\frac{270}{330} =$

Q7 $\frac{90}{245} =$

Q8 $\frac{2205}{7371} =$

Q9 $\frac{286}{663} =$

Q10 $\frac{351}{462} =$

Fraction operations

Adding fractions

Adding fractions with the same denominator

Worked example

To add fractions that share a common denominator, simply add the numerators together.

$$\frac{2}{5} + \frac{1}{5} = \frac{2 + 1}{5} = \frac{3}{5}$$

Adding fractions with different denominators

To add fractions with different denominators, find a common denominator for all the fractions you need to add together. For example:

$$\frac{2}{3} + \frac{4}{5} = ?$$

Method 1: Multiply the denominators to find a common denominator

When the numbers are simple, you can simply multiply the denominators by each other to find a common denominator. Note that this method will not necessarily always give you the lowest common denominator.

$3 \times 5 = 15$, so you know that a common denominator of both fractions is 15.

Tip: If you find it easier to find a common denominator by multiplying the denominators together than to find the lowest common denominator at the start, this approach is fine. Just remember to reduce the fraction to its lowest terms as your final step in the calculation.

Set up the fractions with a common denominator of 15:

$$\frac{?}{15} + \frac{?}{15} = ?$$

To find the equivalent numerators, multiply the numerator of each fraction by the same number as you multiplied the denominator. You multiplied the denominator (3) by 5 in the first fraction to find the common denominator (15), so you must also multiply the numerator (2) by 5.

$2 \times 5 = 10$

Now substitute the numerator into the equation.

$$\frac{10}{15} + \frac{?}{15} = ?$$

Now do the same to the second fraction. You multiplied the denominator (5) by 3 to find a common denominator, so multiply the numerator (4) by 3 as well.

$4 \times 3 = 12$

Substitute the numerator into the equation.

$$\frac{10}{15} + \frac{12}{15} = ?$$

Now you can simply add the numerators:

$$\frac{10}{15} + \frac{12}{15} = \frac{22}{15}$$

The final step is to reduce the fraction to its lowest terms:

$$\frac{22}{15} = 1\frac{7}{15}$$

Method 2: Find the lowest common denominator

First, find the lowest common multiple of the denominators. The lowest common multiple of 6 and 9 is the product of the prime factorization of each number.

$$6 = 2 \times 3$$
$$9 = \quad 3 \times 3$$

(
Common prime factor

The lowest common multiple is therefore $2 \times 3 \times 3 = 18$. Rewrite each of the fractions with a denominator of 18:

$$\frac{1}{6} + \frac{4}{9} = \frac{(1 \times 3)}{18} + \frac{(2 \times 4)}{18} =$$

Now add the numerators:

$$\frac{3}{18} + \frac{8}{18} = \frac{11}{18}$$

The fraction cannot be reduced any further.

Worked example

Adding mixed fractions

A mixed fraction is a fraction consisting of an integer and a fraction. To add mixed fractions, first ensure that the fractions are set up with common denominators and then add the fractions in the usual way.

$$2\frac{1}{4} + 3\frac{7}{9} = ?$$

First convert the mixed fractions to improper fractions:

$$2\frac{1}{4} + 3\frac{7}{9} = \left(\left(\frac{2 \times 4}{4}\right) + \frac{1}{4}\right) + \left(\left(\frac{3 \times 9}{9}\right) + \frac{7}{9}\right) =$$

$$\frac{9}{4} + \frac{34}{9}$$

Now find the lowest common multiple of 4 and 9:

$4 = 2 \times 2$
$9 = 3 \times 3$

There are no common prime factors, so the lowest common multiple is a product of the prime factors: $2 \times 2 \times 3 \times 3 = 36$. Now set up each fraction with a denominator of 36:

$$\frac{(9 \times 9)}{(9 \times 4)} + \frac{(4 \times 34)}{(4 \times 9)} = \frac{81}{36} + \frac{136}{36}$$

Now add the numerators, and reduce the improper fraction to its lowest terms:

$$\frac{81}{36} + \frac{136}{36} = \frac{217}{36} = 6\frac{1}{36}$$

Adding fractions: practice drill

Set a stopwatch and aim to complete the following drill in three minutes.

Q1	$\frac{1}{2} + \frac{3}{4}$	Q2	$\frac{2}{3} + \frac{5}{8}$
Q3	$\frac{1}{8} + \frac{2}{7}$	Q4	$\frac{5}{6} + \frac{7}{9}$
Q5	$3\frac{1}{3} + \frac{1}{5}$	Q6	$\frac{4}{7} + \frac{11}{2}$
Q7	$4\frac{3}{4} + 2\frac{2}{5}$	Q8	$\frac{5}{9} + \frac{2}{7}$
Q9	$\frac{11}{4} + \frac{12}{11}$	Q10	$1\frac{3}{7} + 3\frac{5}{6}$

Subtracting fractions

Subtracting fractions with the same denominator

Worked example

To subtract fractions with the same denominators, simply subtract the numerators.

$$\frac{7}{9} - \frac{5}{9} = \frac{7-5}{9} = \frac{2}{9}$$

Subtracting fractions with different denominators

Worked example

To subtract fractions with different denominators, first find the lowest common denominator.

$$\frac{2}{3} - \frac{4}{5} = ?$$

The lowest common denominator of 3 and 5 = (3 × 5) = 15, so set the fractions up with a common denominator of 15:

$$\frac{10}{15} - \frac{12}{15} = ?$$

Now you can subtract the numerators:

$$\frac{10}{15} - \frac{12}{15} = \frac{10 - 12}{15} = -\frac{2}{15}$$

Subtracting mixed fractions

Worked example

$$1\frac{1}{2} - \frac{2}{3} = ?$$

Convert the mixed fraction to an improper fraction $1\frac{1}{2} = \frac{3}{2}$: Now find the lowest common denominator. The lowest common denominator of 2 and 3 = 2 × 3, so set up the fractions with a common denominator of 6:

$$\frac{3}{2} - \frac{2}{3} = \frac{(3 \times 3)}{6} - \frac{(2 \times 2)}{6}$$

Now subtract the numerators:

$$\frac{9}{6} - \frac{4}{6} = \frac{9 - 4}{6} = \frac{5}{6}$$

The fraction cannot be reduced any further.

Subtracting fractions: practice drill

Set a stopwatch and aim to complete the following drill in two minutes.

Q1 $\frac{4}{5} - \frac{1}{5}$

Q2 $\frac{3}{4} - \frac{1}{4}$

Q3 $\frac{3}{5} - \frac{1}{4}$

Q4 $\frac{5}{6} - \frac{4}{9}$

Q5 $\frac{2}{9} - \frac{1}{12}$

Q6 $\frac{3}{4} - \frac{6}{7}$

Q7 $1\frac{1}{7} - \frac{1}{9}$

Q8 $2\frac{1}{4} - 1\frac{7}{8}$

Q9 $3\frac{1}{3} - 2\frac{2}{11}$

Q10 $2\frac{3}{5} - 3\frac{2}{3}$

Multiplying fractions

Proper fractions

Worked example

To multiply a proper fraction, multiply the numerators together and multiply the denominators together then reduce the result to its lowest terms.

$$\frac{2}{3} \times \frac{4}{5} = \frac{2 \times 4}{3 \times 5} = \frac{8}{15}$$

Mixed fractions

Worked example

$$1\frac{1}{3} \times 1\frac{1}{4} \times \frac{1}{2} = ?$$

To multiply mixed fractions, first convert all the mixed fractions to improper fractions and then multiply the numerators together and the denominators together. Reduce the result to its lowest terms.

$$\frac{4}{3} \times \frac{5}{4} \times \frac{1}{2} = ?$$

You can cancel out equivalent terms before you start the calculation. If you get into the habit of reducing fractions before multiplying, you will minimize errors when you multiply complex fractions and large numbers.

$$\frac{{}^{1}\cancel{4}}{3} \times \frac{5}{\cancel{4}_{1}} \times \frac{1}{2} = ?$$

Now multiply the numerators and the denominators:

$$\frac{1 \times 5 \times 1}{3 \times 1 \times 2} = \frac{5}{6}$$

Multiplying fractions: practice drill

Set a stopwatch and aim to complete the following drill in two minutes.

Q1 $\frac{1}{4} \times \frac{1}{2} =$

Q2 $\frac{1}{3} \times \frac{5}{6} =$

Q3 $\frac{2}{7} \times \frac{3}{4} =$

Q4 $-\frac{1}{8} \times \frac{1}{4} =$

Q5 $\frac{3}{8} \times -\frac{1}{6} =$

Q6 $-\frac{1}{3} \times -\frac{1}{2} =$

Q7 $\frac{1}{2} \times \frac{2}{3} \times \frac{3}{4} =$

Q8 $\frac{5}{6} \times \frac{2}{9} \times \frac{3}{18} =$

Q9 $-\frac{2}{3} \times -\frac{1}{4} \times \frac{3}{12} =$

Q10 $2\frac{1}{3} \times 1\frac{1}{4} \times 3\frac{1}{2} =$

Willesden Green Library

Phone ????
www.bibliotheca-rfid.com

Borrowing

User Mr Paul Lule
User ID 911927

Due Da

t

1 Numerical reasoning tests: 02/06/

0

2 How to master psychometric 02/06/

0

25.05.2009 12:09:

4

Thank you and see you soon!

Willesden Green Library

Phone ????

www.bibliotheca-rfid.com

Returning

1 How to master psychometric tests: exp
e

2 Numerical reasoning tests: a step-by-
s

3 Personality profiler

4 Brilliant Microsoft Access 2007

5 Brilliant Microsoft Word 2007

25.05.2009 12:06:
5

Thank you and see you soon!

Dividing fractions

Proper fractions

To divide two proper fractions, invert the second fraction and multiply the fractions together.

Worked example

$$\frac{1}{2} \div \frac{4}{5} = ?$$

$$\frac{1}{2} \div \frac{4}{5} = \frac{1}{2} \times \frac{5}{4} = \frac{1 \times 5}{2 \times 4} = \frac{5}{8}$$

The result cannot be reduced any further.

Worked example

$$\frac{2}{3} \div \frac{4}{7} = ?$$

$$\frac{2}{3} \div \frac{4}{7} = \frac{2}{3} \times \frac{7}{4} = \frac{2 \times 7}{3 \times 4} = \frac{14}{12}$$

Reduce the fraction to its lowest terms:

$$\frac{14}{12} = 1\frac{1}{6}$$

Dividing fractions: practice drill

Set a stopwatch and aim to complete the following drill in two minutes.

Q1 $\frac{1}{3} \div \frac{2}{5} =$ Q2 $\frac{4}{5} \div \frac{2}{3} =$

Q3 $\frac{3}{7} \div \frac{6}{7} =$ Q4 $\frac{2}{3} \div \frac{4}{9} =$

Q5 $\frac{12}{15} \div \frac{7}{9} =$ Q6 $\frac{1}{9} \div \frac{2}{7} =$

Q7 $2\frac{1}{3} \div 4\frac{1}{4} =$ Q8 $4\frac{3}{4} \div 6\frac{2}{3} =$

Q9 $\dfrac{\frac{5}{6}}{\frac{1}{3}} =$ Q10 $\dfrac{\frac{1}{3} \div \frac{1}{2}}{\frac{1}{4} \div \frac{3}{4}} =$

Decimal operations

Adding decimals

To add decimals, line up the decimal points and add in the usual way.

Worked example

33.75 + 0.46 = ?

$$\begin{array}{r} 33.75 \\ +0.46 \\ \hline 34.21 \end{array}$$

Adding decimals: practice drill

Set a stopwatch and aim to complete the following drill in two minutes.

Q1	27.97 +	46.83 =
Q2	162.72 +	87.03 =
Q3	87.003 +	0.04 =
Q4	34.01 +	152.9 =
Q5	0.001 +	0.002 =
Q6	9.86 +	5.7602 =
Q7	2.561 +	0.002 =
Q8	–1.002 +	1.002 =
Q9	9.87 +	–0.67 =
Q10	–0.01 +	–123.25 =

Subtracting decimals

Subtract decimals by lining up the decimal points and subtracting as normal.

Worked example

$$\begin{array}{r} 4.007 \\ -3.878 \\ \hline 0.129 \end{array}$$

4.007 – 3.878 = ?

Tip: When subtracting numbers, you may find it easier to count up from the number being subtracted. For example, for the subtraction 33.26 – 31.96, count up from 31.96 to 32 to give 0.04. The difference between 32 and 33.26 = 1.26, so the difference between 31.96 and 33.26 = 1.26 + 0.04 = 1.3.

Subtracting decimals: practice drill

Set a stopwatch and aim to complete the following drill in two minutes.

Q1	14.93	–	13.88 =
Q2	12.22	–	11.03 =
Q3	7.003	–	0.041 =
Q4	153.01	–	152.0 =
Q5	1.002	–	1.001 =
Q6	9.16	–	5.76 =
Q7	2.561	–	0.0027 =
Q8	–5.0201	–	4.897 =
Q9	12.80	–	–0.0078 =
Q10	–0.01	–	–0.001 =

Multiplying decimals

To multiply decimals, ignore the decimal points and multiply the numbers as if they were whole numbers. At the end of the calculation, you will reinsert the decimal point.

Worked example

$$\begin{array}{r} 0056 \\ \times\ \ 32 \\ \hline 112 \\ +1680 \\ \hline 1792 \end{array}$$

32 × 0.056

Now count up all the spaces after the decimal points of both original numbers. There are no decimal spaces to consider in

32. In 0.056 there are three spaces after the decimal point, so you must count back three spaces in your final answer and

insert the decimal point.
Answer = 1.792

Worked example

17.6 × 2.1002 =

$$\begin{array}{r} 176 \\ \times \quad 21002 \\ \hline 352 \\ 176000 \\ +3520000 \\ \hline 3696352 \end{array}$$

Line up the numbers, ignoring the decimal points
Count up the decimal places in each number. 17.6 = one decimal place; 2.1002 = four decimal places, so you must count

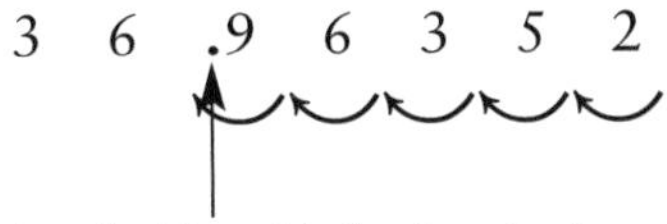

back (4 + 1) decimal places.

Insert the decimal point here.
Answer = 36.96352

Multiplying decimals: practice drill

Set a stopwatch and aim to complete the following drill in two minutes.

Q1	1.92	×	12 =
Q2	12.2	×	11 =
Q3	17.3	×	0.4 =
Q4	15	×	1.4 =
Q5	1.2	×	1.2 =
Q6	19.4	×	5.75 =
Q7	4.51	×	0.271 =
Q8	77.01	×	0.04 =
Q9	13	×	0.0078 =
Q10	0.001	×	0.003 =

Dividing decimals

To divide decimals, treat all the integers in the calculation as if they were whole numbers. To do this, you will increase each integer by a multiple of 10 to produce a whole number.

Worked example

150 ÷ 7.5 = ?

Multiply both numbers by 10.

1500 ÷ 75 = 20

If only one integer is a decimal, you can increase the decimal number by a multiple of 10 and balance the calculation at the end.

Worked example

1000 ÷ 12.5 = ?

Multiply 12.5 by 10.

1000 ÷ 125 = 8

Now multiply the answer by 10 to find the result of 1000 ÷ 12.5 (12.5 is 10 times smaller than 125).

1000 ÷ 12.5 = 80

Dealing with remainders

Worked example

311.6 ÷ 15.2 = ?

Multiply both numbers by a multiple of 10.

(10 × 311.6) ÷ (10 × 15.2) = 3116 ÷ 152

If you find you have a remainder when completing the long division calculation, insert a decimal point and extra zero(s) and continue to divide until either there is no remainder or the calculation produces a recurring number for your answer.

$$152\overline{)3116.0} \quad 20.5$$

Insert decimal point and extra zeros as necessary.

Dividing decimals: practice drill

Set a stopwatch and aim to complete the following drill in two minutes.

Q1	30 ÷ 2.5 =
Q2	90 ÷ 4.5 =
Q3	122.5 ÷ 4.9 =
Q4	326.6 ÷ 14.2 =
Q5	144.5 ÷ 8.5 =
Q6	1.26 ÷ 0.06 =
Q7	17.49 ÷ 2.12 =
Q8	11.4 ÷ 57 =
Q9	27.9 ÷ 4.5 =
Q10	12.88 ÷ 5.6 =

Answers to Chapter 2

Comparing positive fractions: practice drill

Q1 $\frac{2}{3} < \frac{13}{19}$

Q2 $\frac{3}{4} < \frac{13}{16}$

Q3 $\frac{7}{9} > \frac{5}{7}$

Q4 $\frac{3}{9} < \frac{3}{7}$

Q5 $\frac{4}{17} > \frac{2}{9}$

Q6 $\frac{8}{15} < \frac{5}{9}$

Q7 $\frac{28}{51} < \frac{8}{14}$

Q8 $\frac{2}{50} = \frac{7}{175}$

Q9 $\frac{2}{7} = \frac{12}{42}$

Q10 $\frac{100}{4} > \frac{4}{1}$

Reducing fractions: practice drill

Q1 $\frac{6}{7}$

Q2 $\frac{3}{7}$

Q3 $\frac{11}{13}$

Q4 $\frac{5}{21}$

Q5 $\frac{6}{7}$

Q6 $\frac{9}{11}$

Q7 $\frac{18}{49}$

Q8 $\frac{35}{117}$

Q9 $\frac{22}{51}$

Q10 $\frac{117}{154}$

Adding fractions: practice drill

Q1 $1\frac{1}{4}$

Q2 $1\frac{7}{24}$

Q3 $\frac{23}{56}$

Q4 $1\frac{11}{18}$

Q5 $3\frac{8}{15}$

Q6 $6\frac{1}{14}$

Q7 $7\frac{3}{20}$

Q8 $\frac{53}{63}$

Q9 $3\frac{37}{44}$

Q10 $5\frac{11}{42}$

Subtracting fractions: practice drill

Q1 $\frac{3}{5}$

Q2 $\frac{1}{2}$

Q3 $\frac{7}{20}$

Q4 $\frac{7}{18}$

Q5 $\frac{5}{36}$

Q6 $-\frac{3}{28}$

Q7 $1\frac{2}{63}$

Q8 $\frac{3}{8}$

Q9 $1\frac{5}{33}$

Q10 $-1\frac{1}{15}$

Multiplying fractions: practice drill

Q1 $\frac{1}{8}$

Q2 $\frac{5}{18}$

Q3 $\frac{3}{14}$

Q4 $-\frac{1}{32}$

Q5 $-\frac{1}{16}$

Q6 $\frac{1}{6}$

Q7 $\frac{1}{4}$

Q8 $\frac{5}{162}$

Q9 $\frac{1}{24}$

Q10 $10\frac{5}{24}$

Dividing fractions: practice drill

Q1 $\frac{5}{6}$

Q2 $1\frac{1}{5}$

Q3 $\frac{1}{2}$

Q4 $1\frac{1}{2}$

Q5 $1\frac{1}{35}$

Q6 $\frac{7}{18}$

Q7 $\frac{28}{51}$

Q8 $\frac{57}{80}$

Q9 $2\frac{1}{2}$

Q10 2

Adding decimals: practice drill

Q1	74.8
Q2	249.75
Q3	87.043
Q4	186.91
Q5	0.003
Q6	15.6202
Q7	2.563
Q8	0
Q9	9.2
Q10	–123.26

Subtracting decimals: practice drill

Q1	1.05
Q2	1.19
Q3	6.962
Q4	1.01
Q5	0.001
Q6	3.4
Q7	2.5583
Q8	–9.9171
Q9	12.8078
Q10	–0.009

Multiplying decimals: practice drill

Q1	23.04
Q2	134.2
Q3	6.92
Q4	21
Q5	1.44
Q6	111.55
Q7	1.22221
Q8	3.0804
Q9	0.1014
Q10	0.000003

Dividing decimals: practice drill

Q1	12
Q2	20
Q3	25
Q4	23
Q5	17
Q6	21
Q7	8.25
Q8	0.2
Q9	6.2
Q10	2.3

3

Rates

Chapter topics

- Terms used in this chapter
- Converting units
- Working with rates
- Work rate problems
- Answers to Chapter 3

Terms used in this chapter

Rate: A ratio that establishes the relationship between two or more different quantities measured in different units.

Knowledge of rates is very useful for any commercial activity that requires you to measure productivity. A key measure of industrial performance may be the rate at which a finished part is produced, for example the number of surfboards waxed in a day or the number of barrels of oil extracted per month. Worker productivity may be measured by the completed number of manual tasks, such as the number of birthday cakes iced per hour or the number of violins varnished in a week. While these formulae do not account for quality control, you may apply an additional formula to find out, for example, the number of birthday cakes spoilt per week during the icing process.

Converting units

Before we reintroduce the formulae to work out rates questions, a word or two on units. Rates are a useful method to compare different units, but like units must be measured in comparable terms. For example, you cannot always make a useful comparison between 'miles per gallon' and 'kilometres per gallon' or between 'parts produced per worker' and 'parts produced per production line'. The following drills aim to hone your quick conversion skills and will remind you of the common units that you are likely to come across in your test. Where it is critical to know conversion rates, the test will typically contain this information. However, sound knowledge of these common equivalents will help you to calculate your answers quickly and save you those valuable seconds.

Time

Time is the unit that is most frequently converted when working with rates. The basic units of time are seconds, minutes and hours, and you will be required to convert units of time easily between the other time units. The following table is a guide to common time units.

Fraction of hour	Number of minutes
$\frac{1}{10}$ hour	6 minutes
$\frac{2}{10}$ hour	12 minutes
$\frac{3}{10}$ hour	18 minutes
$\frac{4}{10}$ hour	24 minutes
$\frac{5}{10}$ hour	30 minutes
$\frac{6}{10}$ hour	36 minutes
$\frac{7}{10}$ hour	42 minutes
$\frac{8}{10}$ hour	48 minutes
$\frac{9}{10}$ hour	54 minutes
$\frac{1}{3}$ hour	20 minutes
$\frac{2}{3}$ hour	40 minutes
$\frac{1}{6}$ hour	10 minutes
$\frac{5}{6}$ hour	50 minutes
$\frac{1}{4}$ hour	15 minutes
$\frac{3}{4}$ hour	45 minutes

Distance

Distance is another unit that you may be required to convert. The table below gives the common equivalents of metric measures of distance. You will have the opportunity to practise these conversions in the practice drills.

To convert	Equivalents	Calculation
Km to metres	1 km = 1,000 m	Multiply the number of kilometres by 1,000
Km to cm	1 km = 100,000 cm	Multiply the number of kilometres by 100,000
Km to mm	1 km = 1,000,000 mm	Multiply the number of kilometres by 1,000,000
Metres to km	1 m = 0.001 km	Divide the number of metres by 1,000
Metres to cm	1 m = 100 cm	Multiply the number of metres by 100
Metres to mm	1 m = 1,000 mm	Multiply the number of metres by 1,000
Cm to km	1 cm = 0.00001 km	Divide the number of cm by 100,000
Cm to metres	1 cm = 0.01 m	Divide the number of cm by 100
Cm to mm	1 cm = 10 mm	Multiply the number of cm by 10
Mm to km	1 mm = 0.000001 km	Divide the number of mm by 1,000,000
Mm to metres	1 mm = 0.001 m	Divide the number of mm by 1,000
Mm to cm	1 mm = 0.1 cm	Divide the number of mm by 10

Common units

To convert...	Equivalents	Calculation
Years to weeks	1 year = 52 weeks	Multiply the number of years by 52
Weeks to days	1 week = 7 days	Multiply the number of weeks by 7
Weeks to hours	1 week = 168 hours	Multiply the number of weeks by 168
Weeks to minutes	1 week = 10,080 minutes	Multiply the number of weeks by 10,080
Days to hours	1 day = 24 hours	Multiply the number of days by 24
Hours to minutes	1 hour = 60 minutes	Multiply the number of hours by 60
Hours to seconds	1 hour = 3,600 seconds	Multiply the number of hours by 3,600
Minutes to seconds	1 minute = 60 seconds	Multiply the number of minutes by 60
Feet to inches	1 foot = 12 inches	Multiply the number of feet by 12
Miles to km	1 mile = approx. 1.61 km	Multiply the number of miles by 1.61
Km to miles	1km = approx. 0.62 mile	Multiply the number of km by 0.62
Ounces to grammes	1 oz = 25 grammes	Multiply the number of ounces by 25
Inches to cm	1 inch = 2.5cm	Multiply the number of inches by 2.5
Pints to millilitres	1 pint = 570 ml	Multiply the number of pints by 570
Farenheit to Celcius	(F – 32) × 0.55	Subtract 32 from the number of degrees F and multiply by 0.55
Celcius to Farenheit	(C × 1.8) + 32	Multiply the number of degrees C by 1.8 and add 32

Rates conversion practice: drill 1: time

Set a stopwatch and aim to complete the following drill in three minutes.

Q1	$2\frac{1}{4}$ hours	=	minutes
Q2	$6\frac{3}{4}$ hours	=	minutes
Q3	$10\frac{1}{3}$ hours	=	minutes
Q4	$1\frac{5}{6}$ hours	=	minutes
Q5	$\frac{1}{6}$ hour	=	minutes
Q6	480 minutes	=	hours
Q7	320 minutes	=	hours
Q8	280 minutes	=	hours
Q9	50 minutes	=	hour
Q10	144 minutes	=	hours
Q11	37 minutes	=	seconds
Q12	36 seconds	=	hour
Q13	$\frac{1}{10}$ hour	=	seconds
Q14	$12\frac{1}{4}$ hours	=	seconds
Q15	48 minutes	=	hour
Q16	$\frac{3}{10}$ hour	=	seconds
Q17	$\frac{1}{4}$ hour	=	seconds
Q18	20 seconds	=	minute
Q19	$\frac{7}{10}$ hour	=	minutes
Q20	0.2 hour	=	minutes

Rates conversion: practice drill 2: distance

Set a stopwatch and aim to complete the following drill in three minutes.

Q1	25mm	=	metres
Q2	2 cm	=	km
Q3	72 inches	=	feet
Q4	1598 cm	=	km

Q5	¼ inch	=	foot
Q6	956 cm	=	km
Q7	12.91 cm	=	metres
Q8	83 cm	=	mm
Q9	39 inches	=	feet
Q10	6 mm	=	km

Rates conversion: practice drill 3: temperature

Set a stopwatch and aim to complete the following drill in three minutes. Convert the following temperatures to the nearest whole number:

Q1	32°C	=	°F
Q2	37.5 °C	=	°F
Q3	3°C	=	°F
Q4	572°C	=	°F
Q5	0.25°C	=	°F
Q6	99.68°F	=	°C
Q7	5°F	=	°C
Q8	8°F	=	°C
Q9	98.6°F	=	°C
Q10	105°F	=	°C

Working with rates

A rate is a ratio, which establishes the relationship between two or more different quantities measured in different units. For example, you are probably familiar with the frequently used rates of 'miles per gallon' or 'miles per hour'. Rates are used to measure proportions between different units and are a useful method to compare quantities of the different units, for example 'cost per square metre'. This chapter will demonstrate how to substitute the variables into a common formula. The examples use time, speed and distance to demonstrate the concept. Hints on how to recognize other variables you may come across in the test are also provided.

There are three parts to a speed-distance-time problem. The key to answering these questions correctly is to identify the pieces correctly and substitute the information to the relevant formulae. The relationship between the three pieces of the puzzle is expressed as:

$$\text{Distance} = \text{Rate} \times \text{Time}$$

This can be rearranged as:

$$\text{Time} = \frac{\text{Distance}}{\text{Rate}}$$

Or in other words:

$$\text{Rate} = \frac{\text{Distance}}{\text{Time}}$$

The following sections work through examples of the use of each of the formulae.

Find the distance when you know the time and rate

Distance = Rate × Time

Worked example

A car travels along a country road at a rate of 45 mph for 5½ hours. How far does the car travel?

Estimate the answer

If the car travelled at 40 mph for 5 hours, the distance travelled would be:

40 mph × 5 hours = 200 miles.

As the actual speed is quicker and the time spent travelling longer, your answer will be slightly higher than 200.

Calculate the answer

Apply the formula to find the distance:

Rate	×	**Time =**	**Distance**
45mph	×	5.5 hours =	247.5 miles

Distance travelled = 247.5 miles.

Worked example

Ruth sets out to paddle the Colorado River through the Grand Canyon. She paddles for 2 hours at a rate of 8.25 miles per hour while the river is calm. On approaching her first set of rapids, her speed increases to 18.5 miles per hour for half an hour until she reaches her first waterfall. How far is Ruth from her launch point when she reaches her first waterfall?

Estimate the answer

There are two speeds to consider here, and therefore you will calculate the distance in two stages. If Ruth travelled for 2 hours at 8 mph, she would travel 16 miles. If she travelled a further hour at 18 mph, she would travel 18 miles, so in half an hour she would travel for 9 miles. The total distance travelled therefore is approximately 16 miles + 9 miles = 25 miles.

Calculate the answer

Apply the formula to find the distance for each leg of the journey

Rate	×	Time =	Distance
8.25mph	×	2 hours =	16.5 miles
18.5mph	×	0.5 hours =	9.25 miles

Total distance travelled = $16\frac{1}{2} + 9\frac{1}{4} = 25\frac{3}{4}$ miles.

Find the time when you know the distance and rate

$$\text{Time} = \frac{\text{Distance}}{\text{Rate}}$$

Worked example

Iain lives on a small island off the West Coast of Scotland and rows his boat to the mainland once a week to collect provisions. The distance from the mooring point on the island to the mooring point on the mainland is half a mile. He rows his boat at a rate of 2 miles an hour when the wind is still. How many minutes does it take Iain to row to the mainland on a calm day?

Estimate the answer

If Iain rows for an hour, he will travel 2 miles. Since you know that the total distance he has to travel is less than 2 miles, you know that he will not be rowing for a whole hour.

Calculate the answer

Apply the formula and plug in the numbers:

$$\text{Time} = \frac{\text{Distance}}{\text{Rate}} = \frac{0.5 \text{ miles}}{2 \text{ mph}} = \frac{\not{5}}{\not{20}}\ \frac{1}{4}$$

It takes Iain $\frac{1}{4}$ of an hour, or 15 minutes, to row to the mainland on a calm day.

Worked example

Captain Bengel is a jump-jet pilot based in the desert in Arizona. On one particular exercise, Bengel and his wingman, Captain Danger, set off from their airbase in Arizona to fly 320 miles to Las Vegas in Nevada at a constant average speed of 800 mph. On the return journey, the pilots are forced to reduce their speed to conserve fuel and return at a rate of 400 mph. What is the total flying time for both pilots together in hours and minutes, assuming simultaneous take-off and landing?

Estimate the answer

At a speed of 800 mph, the pilots would travel 800 miles in an hour, so the outbound journey of less than 320 miles will take a little less than half an hour. On the return journey of 320 miles at a speed of 400 mph, the journey would take a little less than an hour. The total flying time will therefore be (approximately $\frac{1}{2}$ hour) + (approximately 1 hour) = approximately $1\frac{1}{2}$ hours.

Calculate the answer

Apply the formula to find the time for both the outbound and inbound journeys.

Outbound journey:

$$\text{Time} = \frac{\text{Distance}}{\text{Rate}} = \frac{320 \text{ miles}}{800 \text{ mph}}$$

Reduce the fraction to its lowest terms:

$$\frac{320\text{ miles}}{800\text{ mph}} = \frac{4}{10}$$

$$\frac{4}{10} \times 60\text{ minutes} = 24\text{ minutes}$$

Inbound journey:

$$\text{Time} = \frac{\text{Distance}}{\text{Rate}} = \frac{320\text{ miles}}{400\text{ mph}}$$

Reduce the fraction to its lowest terms:

$$\frac{320\text{ miles}}{400\text{ mph}} = \frac{4}{5}$$

$$\frac{4}{5} \times 60\text{ minutes} = 48\text{ minutes}$$

Total flying time for each pilot = 24 minutes + 48 minutes = 72 minutes. So the total time for both pilots is 144 minutes or 2 hours and 24 minutes.

Find the rate when you know the distance and time

$$\text{Rate} = \frac{\text{Distance}}{\text{Time}}$$

Often, questions concerning rates relate to speed and you will need to apply this formula to find the answer to 'find the speed' type questions.

Worked example

Sanjay sets off walking to work at 7.10 am. He stops to buy a coffee and read his newspaper for 15 minutes and arrives at work at 8.55 am. The distance between his home and work is 6 miles. What is his average walking speed?

Estimate the answer

The actual walking time is 1 hour and 30 minutes and the total distance covered is 6 miles. If the time spent walking were 1 hour, the rate would be 6 mph. As the time taken to walk the distance is longer than 1 hour, the rate is slower. Your answer will therefore be less than 6 mph.

Calculate the answer

$$\text{Rate} = \frac{\text{Distance}}{\text{Time}}$$

First convert the time to minutes. 1 hour 30 minutes = 90 minutes. Now apply the formula and plug in the numbers.

$$\text{Rate} = \frac{\text{Distance}}{\text{Time}} = \frac{6 \text{ miles}}{90 \text{ minutes}} = \frac{1 \text{ mile}}{15 \text{ minutes}}$$

If Sanjay walks 1 mile in 15 minutes, he will walk 4 miles in 1 hour. The rate is therefore 4 mph.

Worked example: finding the average of two speeds

Jamie enters a biathlon, which requires her first to ride her bike between Thorpe Bay and Chalk Park and then to turn around and run back to Thorpe Bay on the same path. In this particular biathlon, Jamie cycles at a speed of 12 mph and runs at a speed of 6 mph. What is her average speed for the race?

Estimate the answer

When you combine average speeds, you cannot simply add the two together and then divide by 2, as you would do to find an average of two numbers. You can assume that the average speed will be closer to the lower of the two numbers (ie 6 mph) as more time will be spent on the leg of the journey that has the lower rate.

Calculate the answer

The formula to find the average rate of two (or more) rates is:

$$\text{Average rate} = \frac{\text{Total distance}}{\text{Total time}}$$

As you do not know the distance Jamie will travel, pick a distance that creates an easy calculation. For example, since 6 and 12 are both multiples of 24, you could choose 24 miles as the distance between the two points. (Equally, you could choose 12.) It may help you to draw a diagram to facilitate your thinking.

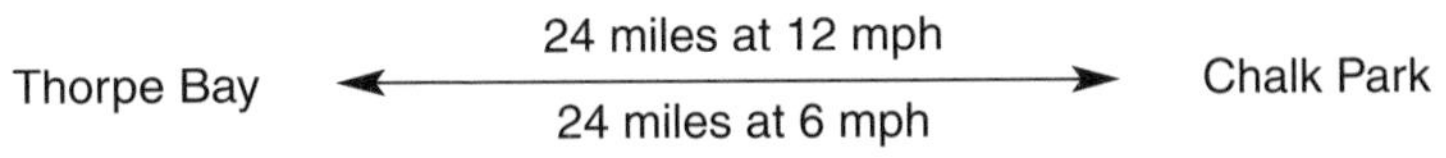

First work out the time it takes Jamie to complete each leg of the journey by plugging in the numbers to the relevant formula. In this instance, use:

$$\text{Time} = \frac{\text{Distance}}{\text{Rate}}$$

Outbound time: Thorpe Bay to Chalk Park:

$$\text{Time} = \frac{\text{Distance}}{\text{Rate}} = \frac{24 \text{ miles}}{12 \text{ mph}} = 2 \text{ hours}$$

Inbound time: Chalk Park to Thorpe Bay:

$$\text{Time} = \frac{\text{Distance}}{\text{Rate}} = \frac{24 \text{ miles}}{6 \text{ mph}} = 4 \text{ hours}$$

So it takes Jamie (2 + 4) hours to complete the whole distance of 48 miles. Now apply the formula to find the average rate.

$$\text{Average rate} = \frac{\text{Total distance}}{\text{Total time}} = \frac{48 \text{ miles}}{6 \text{ hours}} = 8 \text{ mph}$$

So Jamie's average speed for the whole distance is 8 mph.

Summary of rates formulae

Distance = Rate × Time

$$\textbf{Time} = \frac{\text{Distance}}{\text{Rate}}$$

$$\textbf{Rate} = \frac{\text{Distance}}{\text{Time}}$$

Rates practice questions

Set a stopwatch and aim to complete each question within 80 seconds.

Q1 Yousuf sets off on the annual London to Cambridge bike ride. He cycles at a steady pace of 9 miles an hour and crosses the finishing line 6 hours later. What is the distance of the London to Cambridge ride?

Q2 Jo walks along the South Downs Way at a rate of 4.5 miles per hour. After walking for 3½ hours, she still has 6 miles left to walk. What is the total distance she intends to walk?

Q3 A singer is on stage with a bass guitarist. Part of the dance routine requires them to move to opposite ends of the stage for one section of the song. Standing together, they start to move away from each other. The singer shimmies across the stage at the speed of 1 foot every 10 seconds, while the bassist shuffles at the speed of 1 foot every 20 seconds. After 2 minutes, they both reach the edges of the stage. How wide is the stage?

Q4 A rowing boat travels at an average speed of $\frac{1}{1000}$ mile every second. How far will the boat travel in 2 hours?

Q5 A farmer leaves a bale of hay every morning at a location on Dartmoor for the wild ponies, while another farmer leaves a bucket of oats at another location in the evening. One pony starts at the hay and trots along a track to the location of the oats at a rate of 12 mph and remains at that location for the night. In the morning, he covers the same distance back to the hay at a rate of 6 mph. His round trip journey takes exactly 2 hours and he always starts the journey at the location of the hay. What is the distance between the location of the hay and the location of the oats?

Q6 On a particular stretch of railway line, a speed restriction of 50 per cent of the maximum speed of 100 mph is imposed on parts of the track. What is the minimum time a traveller must allow to complete a journey of 375 miles?

Q7 A hosepipe discharges water at a rate of 4 gallons per minute. How long, in hours, does it take to fill a paddling pool with a 300-gallon capacity?

Q8 Mike drives from his flat to the seaside in 1½ hours. On the way home he is in a hurry to watch the start of the Crystal Palace match and drives one and a half times as fast along the same route. How much time does he spend driving?

Q9 Two cars, 100 miles apart, set off driving towards each other. One is travelling at 70 mph and the other is travelling at 80 mph. After how long will the two cars meet?

Q10 A small car with an engine size of 848 cc and a family car with an engine size of 3 litres set off on a journey of 480 miles. The family car completes the journey in 8 hours. The small car travels at an average speed of ⅚ of the speed of the family car. How long after the family car will the small car arrive at the destination?

Q11 The perimeter of Jenny's local park measures 5.5 miles and it takes Jenny 3 hours to run four times around the park. What is her average speed?

Q12 Jake rides his new tricycle for half an hour in the park and covers ¾ mile. What is his average speed in miles per hour?

Q13 Bob drives a cab in London. He picks up a passenger at Gatwick Airport on a Friday afternoon and drives 26 miles to a Central London location, a journey of 2 hours and 15 minutes. Then he drives an additional 4 miles to Waterloo station, which takes another 45 minutes. What is Bob's average speed for the Friday afternoon journeys?

Q14 An egg timer contains 18,000 grains of sand. The sand passes at a consistent rate between the top and the bottom cylinders. After 2 minutes and 24 seconds, 14,400 grains of sand have passed through the middle. At what rate per minute does the sand pass through the timer?

Q15 In the Ironman Triathlon, athletes are expected to swim for 2.4 miles, cycle for 112 miles and run a full marathon distance of 26.2 miles. Justin enters the competition and is given the following split times for each leg of the competition:

Swim: 48 minutes Cycle: 4 hours 59 minutes
Run: 6 hours 43 minutes

To the nearest miles per hour, what is Justin's average speed for the triathlon?

Work rate problems

Work rate problems require that you work out the time involved to complete a specified number of jobs by a specified number of operators. This section shows you two methods to tackle these questions. Choose the method that is easiest for you, and use the second method to check your answer.

Worked example

Work rate formula 1

To find the combined time of two operators working together but independently on the same job at different rates, use the following formula:

$$\text{Time combined} = \frac{xy}{x + y}$$

If it takes Sue 2 hours to weed the garden and it takes Dave 4 hours to do the same task, how long does it take Sue and Dave to weed the garden when they work on the task together at their individual rates? Apply the formula and plug in the numbers:

$$\text{Time combined} = \frac{2 \times 4 = 8}{2 + 4 = 6}$$

The task will take $\frac{8}{6}$ of an hour, or 1 hour 20 minutes.

Work rate formula 2

The total time required to complete a task by more than one operator is equivalent to the following formula:

$$\frac{1}{T1} + \frac{1}{T2} + \frac{1}{T3} = \frac{1}{T}$$

Worked example

If it takes Sue 2 hours to weed the garden and it takes Dave 4 hours to do the same task, how long does it take Sue and Dave to weed the garden when they work on the task together at their individual rates?

Apply the formula and plug in the numbers:

$$\frac{1}{2} + \frac{1}{4} = \frac{1}{T}$$

$$\frac{3}{4} = \frac{1}{T} \text{ so } \frac{4}{3} = \frac{T}{1}$$

$\frac{4}{3} \times 60$ minutes = 100 minutes, or 1 hour 20 minutes.

Work rates practice questions

Q1 Lisa paints a room in 2 hours and Amanda paints the same room in 3 hours. Working together but independently, how long does it take them to paint the room?

Q2 If Jake can clean his sister's bike in 15 minutes and Lauren, Jake's sister, can clean her own bike in 30 minutes, how many minutes will it take both of them together to clean Lauren's bike?

Q3 Computer A can run a set of tasks in 1 hour while Computer B runs the same set of tasks in 1½ hours. How long will it take both computers working together to run half the set of tasks?

Answers to Chapter 3

Rates conversion: practice drill 1: time

Q1	135 minutes
Q2	405 minutes
Q3	620 minutes
Q4	110 minutes
Q5	10 minutes
Q6	8 hours
Q7	$5\frac{1}{3}$ hours
Q8	$4\frac{2}{3}$ hours
Q9	$\frac{5}{6}$ hour
Q10	$2\frac{2}{5}$ hours
Q11	2,220 seconds
Q12	0.01 hour
Q13	360 seconds
Q14	44,100 seconds
Q15	$\frac{4}{5}$ hour
Q16	1,080 seconds
Q17	900 seconds
Q18	$\frac{1}{3}$ minute
Q19	42 minutes
Q20	12 minutes

Rates conversion: practice drill 2: distance

Q1	0.025 metres
Q2	0.00002 km
Q3	6 feet
Q4	0.01598 km
Q5	$\frac{1}{48}$ foot
Q6	0.00956 km
Q7	0.1291 metres
Q8	830 mm

Q9 3¼ feet
Q10 0.000006 km

Rates conversion: practice drill 3: temperature

Q1 90°F
Q2 100°F
Q3 37°F
Q4 1,062°F
Q5 32°F
Q6 37°C
Q7 –15°C
Q8 –13°C
Q9 37°C
Q10 40°C

Rates practice questions answers (for explanations see overleaf)

Q1 54 miles
Q2 21¾ miles
Q3 18 feet
Q4 7.2 miles
Q5 8 miles
Q6 7½ hours
Q7 1 hour 15 minutes
Q8 2 hours 30 minutes
Q9 40 minutes
Q10 1 hour 36 minutes
Q11 7⅓ mph
Q12 1½ mph
Q13 10 mph
Q14 6,000 grains per minute
Q15 11 mph

Rates practice questions: explanations

Q1 Answer = 54 miles

Estimate the answer

At 10 mph for 6 hours, he would cover 60 miles, so the total distance will be slightly less than 60 miles.

Calculate the answer

Apply the formula to find the distance:

Distance	=	**Rate**	×	**Time**
Distance	=	9 mph	×	6 hours
Distance	=	54 miles		

Q2 Answer = 21¾ miles

Estimate the answer

If she walked for 3 hours at 4 mph, she would complete 12 miles and would still have another 6 to walk. The total distance will be more than (12 + 6) miles = 18 miles as she is walking faster than 4 mph and for longer than 3 hours.

Calculate the answer

Apply the formula and plug in the numbers:

Distance	=	**Rate**	×	**Time**
Distance	=	4.5 mph	×	3.5 hours

Distance walked so far = 15.75 miles.

She still has 6 miles to walk, so add this to the distance already covered = 21¾ miles.

Q3 Answer = 18 feet

Estimate the answer
The singer shimmies 6 feet per minute while the bassist shuffles at half the speed, so he will cover 3 feet in a minute. In two minutes they will cover 2 × (6 feet + 3 feet) = 18 feet.

Calculate the answer
You can work out the rate for one of the two to traverse the whole stage by finding the average speed for both performers:

6 feet per minute + 3 feet per minute = 9 feet per minute

Now apply the formula to find the distance:

Distance	=	**Rate**	×	**Time**
Distance	=	9 feet per minute	×	2 minutes

The stage is 18 feet wide.

Q4 Answer = 7.2 miles

Estimate the answer
The calculation is simple, so move straight on from the estimation.

Calculate the answer
In 1 second, the boat travels $\frac{1}{1000}$ mile.
In 1 hour the boat travels:

$$\frac{1}{1000} \times 3600 \text{ seconds} = 3.6 \text{ miles}$$

In 2 hours the boat travels 2 × 3.6 miles = 7.2 miles.

Q5 Answer = 8 miles

Calculate the answer

You can solve this problem with logic. Set up a diagram to help you to visualize the scenario.

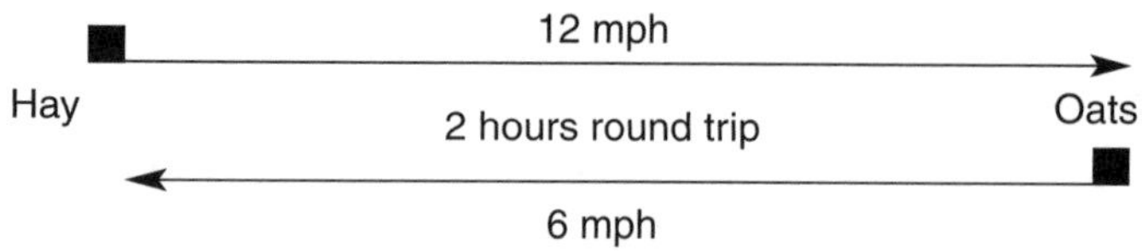

Logic tells you that the return journey will take twice as long as the outbound journey because the pony is travelling at half the speed. In order to set up a ratio, you need to work out how much time each leg of the journey takes. You can set up an equation to help you to solve this. Let's say the outbound journey from Hay to Oats takes M minutes. The return journey will take 2M minutes. Therefore, M + 2M = 2 hours or 3M = 120 minutes. To find M, divide by 3:

$$M = \frac{120}{3} = 40 \text{ minutes}$$

So the outbound journey (M minutes) takes 40 minutes and the return journey (2M minutes) takes 80 minutes (or 2 hours – 40 minutes = 80 minutes). Now you can plug in the numbers to the relevant formulae:

Distance = Time × Rate
Distance = 40 minutes × 12 mph
Distance = ⅔ hour × 12 mph

The distance between the hay and the oats is 8 miles.

Q6 Answer = 7½ hours

Estimate the answer

As we do not know the distance of the restricted stretches of railway, we have to assume that the restriction could apply to the whole journey. If the train travels at 50 per cent of 100 mph, it travels at 50 mph for 375 miles. 375 ÷ 50 = approximately 7, so you are looking for an answer of a little more than 7 hours.

Calculate the answer

Apply the formula for Time and plug in the numbers:

$$\text{Time} = \frac{\text{Distance}}{\text{Rate}} = \frac{375 \text{ miles}}{50 \text{ mph}} = 7\tfrac{1}{2} \text{ hours}$$

The minimum time the traveller must allow is 7½ hours.

Q7 Answer = 1¼ hours

Estimate the answer

In one hour, 4 gallons × 60 minutes water will be discharged, which is 240 gallons. Therefore your answer will be at least one hour and less than two.

Calculate the answer

You are looking for the time required to fill the paddling pool, so set up the formula for Time.

$$\text{Time} = \frac{\text{Distance}}{\text{Rate}} = \frac{300 \text{ gallons}}{4 \text{ gallons per minute}}$$

It takes 75 minutes to fill the paddling pool, but the question asks for the answer in hours, so convert from minutes to hours by dividing by 60:

$75 \div 60 = 1.25$ hours.

Q8 Answer = 2 hours 30 minutes

Estimate the answer

If Mike drove home at the same speed, the total driving time would be 2 × 90 minutes = 180 minutes or 3 hours. If Mike drove twice as fast on the way home, the return journey would take him 45 minutes, or half the time of the outward journey. The total time would be 1½ hours + 45 minutes = 2 hours and 15 minutes. Your answer will therefore be between 2 hours and 15 minutes and 3 hours.

Calculate the answer

You may find it helpful to sketch a diagram to help you visualize the problem.

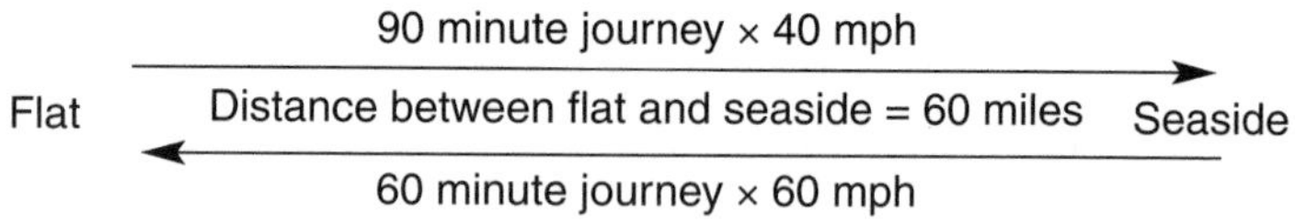

Pick numbers to help you to answer this question. Let's assume that the distance driven is 60 miles. Apply the formula to find the rate:

$$\text{Rate} = \frac{\text{Distance}}{\text{Time}} = \frac{60 \text{ miles}}{1.5 \text{ hours}} = 40 \text{ mph}$$

If he drives one and a half times faster on the way home, he drives at 1.5 × 40 mph = 60 mph. So on the return journey he drives 60 miles at 60 mph and the journey takes 1 hour. The total time is therefore 1 hour 30 minutes + 1 hour = 2 hours and 30 minutes.

Q9 Answer = 40 minutes

Estimate the answer

You may find it helpful to set up a rough diagram to help you visualize what is going on in the question:

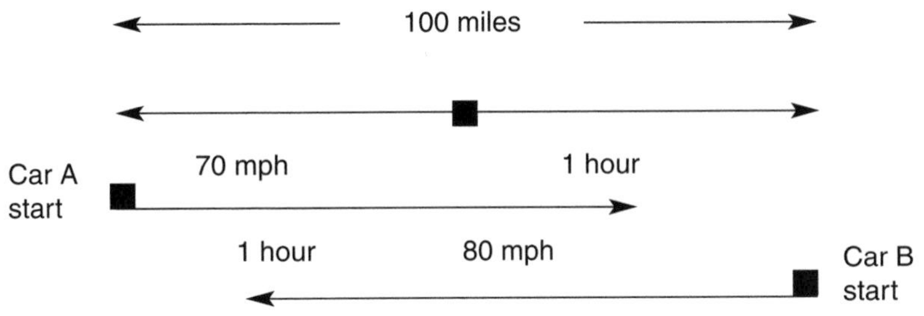

You will see that within an hour, both cars will have travelled more than halfway, and therefore they will have met within 1 hour.

Calculate the answer

If Car A remained stationary and Car B travelled at the combined speeds of Car A and Car B, the meeting point would be the same as if both cars travelled towards each other, setting off at the same time. The combined speed of Car A and Car B = 70 + 80 = 150 mph. Set up the relevant formula to find the time.

$$\text{Time} = \frac{\text{Distance}}{\text{Rate}} = \frac{100 \text{ miles}}{150 \text{ mph}} = \frac{2}{3}$$

$\frac{2}{3}$ hour = 40 minutes. The cars will meet after 40 minutes.

Q10 Answer = 1 hour 36 minutes

Estimate the answer

As the small car is travelling almost as fast as the large car, the small car will arrive later than 8 hours after the start of the journey.

Calculate the answer

In order to find the average speed of the small car, find the average speed of the family car first. Apply the formula to find the rate and plug in the numbers.

$$\text{Rate} = \frac{\text{Distance}}{\text{Time}} = \frac{480 \text{ miles}}{8 \text{ hours}}$$

The family car travels at an average speed of 60 mph. The small car travels at $\frac{5}{6}$ the speed of the family car: $\frac{5}{6} \times 60$ mph = 50 mph. Now apply the formula to find time:

$$\text{Time} = \frac{\text{Distance}}{\text{Rate}} = \frac{480 \text{ miles}}{50 \text{ mph}} = 9.6 \text{ hours}$$

The small car arrives after 9.6 hours. The family car journey takes 8 hours and the small car journey takes 9.6 hours, so the small car arrives 1 hour and 36 minutes after the family car.

Q11 Answer = $7\frac{1}{3}$ mph

Estimate the answer

If Jenny took 4 hours to complete four circuits, you would know that her average speed is 5.5 mph. Since she completes four circuits in less time, her average speed will be slightly faster than 5.5 mph.

Calculate the answer

Find the total distance Jenny runs in four laps of the park: $4 \times 5.5 = 22$ miles. Now apply the formula to find the rate:

$$\text{Rate} = \frac{\text{Distance}}{\text{Time}} = \frac{22 \text{ miles}}{3 \text{ hours}} = 7\tfrac{1}{3} \text{ mph}$$

Jenny's average speed is $7\frac{1}{3}$ mph.

Q12 Answer = $1\frac{1}{2}$ mph

Estimate the answer

The calculation is simple so move straight on and work out the answer.

Calculate the answer

If Jake managed to pedal for another 30 minutes at the same speed, he would pedal $2 \times \frac{3}{4}$ mile = $1\frac{1}{2}$ miles. His speed is therefore $1\frac{1}{2}$ miles per hour.

Q13 Answer = 10 mph

Estimate the answer

The calculation is simple so move straight on and work out the answer.

Calculate the answer

Apply the formula to find the average of two rates and plug in the numbers:

$$\text{Average rate} = \frac{\text{Total distance}}{\text{Total time}} = \frac{(26 + 4) = 30 \text{ miles}}{(2 \text{ hrs } 15 \text{ mins} + 45 \text{ mins}) = 3 \text{ hrs}}$$

The average speed is $\frac{30 \text{ miles}}{3 \text{ hours}} = 10$ mph

Q14 Answer = 6000 grains per minute

Estimate the answer

A quick conversion tells you that 2 minutes and 24 seconds = 144 seconds. If 14,400 grains of sand pass through the timer in 144 seconds, then 14,400/144 grains of sand will pass through every second.

Calculate the answer

Set up a proportion to find the fraction of the sand that has passed through the timer after 2 minutes and 24 seconds. This will allow you to find the total time required for 18,000 grains of sand to pass through.

$\frac{4}{5}$ = 2 minutes 24 seconds. Therefore

$\frac{1}{5}$ = 36 seconds and therefore

$\frac{5}{5}$ = 3 minutes

Now you can set up the formula to find the rate and plug in the numbers:

$$\text{Rate} = \frac{18{,}000 \text{ grains}}{3 \text{ minutes}}$$

Answer = 6,000 grains per minute.

Q15 Answer = 11 mph

Estimate the answer

As there are three different speeds to calculate, work out each one separately.

Calculate the answer

Remember that you cannot simply add the rates together and find the average. First convert the time to the lowest unit (ie minutes):

Swim = 48, Cycle = 299, Run = 403: total = 750 minutes = 12.5 hours.

Now apply the formula to find the average rate and plug in the numbers.

140.6 ÷ 12.5 = 11.248. The question asks for the answer to the nearest mile, so the answer is 11 mph.

$$\text{Average rate} = \frac{\text{Total distance}}{\text{Total time}} = \frac{140.6 \text{ miles}}{12.5 \text{ hours}}$$

Work rates answers

Q1 1 hour 12 minutes
Q2 10 minutes
Q3 18 minutes

Work rates practice questions explanations

Q1 Answer = 1 hour and 12 minutes

Apply the formula and plug in the numbers.

$$\text{Time combined} = \frac{2 \times 3}{2 + 3} = \frac{6}{5}$$

$\frac{6}{5}$ of an hour = 1 hour and 12 minutes.

Q2 Answer = 10 minutes

Apply the formula, plug in the numbers and work through the method.

$$\frac{1}{15} + \frac{1}{30} = \frac{1}{T}$$

Find the lowest common denominator:

$$\frac{2}{30} + \frac{1}{30} = \frac{1}{T}$$

$$\frac{3}{30} = \frac{1}{\mathrm{T}}$$

Reduce to its lowest terms.

$$\frac{1}{10} = \frac{1}{T}$$

Find the reciprocal by inverting the fraction. T = 10 minutes.

Q3 Answer = 18 minutes

First work out how long it will take both computers working together to complete all the tasks by applying the formula and plugging in the numbers.

$$\frac{1}{60} + \frac{1}{90} = \frac{1}{T}$$

Find the lowest common denominator and add the fractions:

$$\frac{3}{180} + \frac{2}{180} = \frac{5}{180} = \frac{1}{T}$$

Reduce the fraction to its lowest terms:

$$\frac{1}{36} = \frac{1}{T}$$

Find the reciprocal or T by inverting the fraction. T = 36 minutes. If it takes 36 minutes to complete a set of tasks, it will take 18 minutes to complete half the set.

4

Percentages

Chapter topics

- Terms used in this chapter
- Converting between percentages, fractions and decimals
- Converter tables
- Working with percentages
- Simple interest and compound interest
- Answers to Chapter 4

Terms used in this chapter

Percentage: One part in every hundred.

A percentage is a special type of ratio, which compares a number to 100. 'Per cent' or the '%' sign mean 'out of 100'. Expressed as a fraction, 1 per cent is expressed as 1 out of 100, or $\frac{1}{100}$, and is the mathematical relationship used most commonly in everyday life. For example, we are used to hearing that our local supermarket has a 10 per cent weekend sale, or that interest rates will change by $\frac{1}{4}$ per cent, or that inflation has risen by 5 per cent on last year's index. The key to understanding problems involving percentages is to ask yourself first what the problem is about, then to decide whether you have enough information to answer the question. This chapter will guide you through the concepts and formulae to help you tackle percentage problems.

Test-writers love percentages! In most aptitude tests you will have to work out percentages, so it is worth understanding how they work. Practise as much as you can, so that you start to recognize the kind of questions you may be asked. Many percentage questions require you to interpret data from tables, charts and graphs. This chapter provides a reminder of the background to percentages, and drills you in word problems involving percentages. In Chapter 6, you will have the opportunity to apply your knowledge of percentages to data presented in a variety of formats.

Converting between percentages, fractions and decimals

When you work with percentages, it is useful to know the decimal and fraction equivalents so that you have a number of tools at your disposal to help you arrive at the correct answer. This section shows you how to convert between percentages, fractions and decimals.

To express a fraction as a percentage

To express ½ as a percentage, multiply by 100:

$$\frac{1}{2} \times \frac{100}{1} = \frac{100}{2}$$

Reduce the fraction to its lowest terms:

$$\frac{100}{2} = \frac{50}{1} \text{ or } 50\%$$

To express ${}^{12}\!/_{4}$ as a percentage, multiply by 100:

$$\frac{12}{4} \times \frac{100}{1} = \frac{1200}{4}$$

Reduce the fraction to its lowest terms:

$$\frac{1200}{4} = \frac{300}{1} \text{ or } 300\%$$

To express a decimal as a percentage

Move the decimal point two places to the right and add the % sign. This is the same as multiplying by 100.

0.025 = 2.5%
1.875 = 187.5%

To express a percentage as a decimal

Drop the % sign and move the decimal point two places to the left. This is the same as dividing by 100.

37.5% = 0.375
253% = 2.53

To express a fraction as a decimal

Divide the numerator by the denominator:

$^1/_4 = 1 \div 4 = 0.25$
$^5/_2 = 5 \div 2 = 2.5$

To express a percentage as a fraction

Remember that a percentage is a fraction of 100, so write the percentage as a fraction where the denominator is 100.

$$24\% = \frac{24}{100} = \frac{6}{25}$$

$$15\% = \frac{15}{100} = \frac{3}{20}$$

To express a decimal as a fraction

Express the decimal as a fraction over 100 and reduce the fraction to its lowest terms:

$$0.25 = \frac{25}{100} = \frac{1}{4}$$

$$2.875 = \frac{287.5}{100} = 2\frac{7}{8}$$

Converter tables

The following tables are a quick reference to percentage, fraction and decimal equivalents. Become familiar with equivalents to save time in your test.

Percentage	Decimal	Fraction
100%	1	$^1/_1$
10%	0.1	$^1/_{10}$
1%	0.01	$^1/_{100}$
0.1%	0.001	$^1/_{1,000}$
0.01%	0.0001	$^1/_{10,000}$
0.001%	0.00001	$^1/_{100,000}$

Percentage	Decimal	Fraction
$\frac{1}{2}$%	0.005	$\frac{1}{200}$
$\frac{1}{4}$%	0.0025	$\frac{1}{400}$
$\frac{1}{8}$%	0.00125	$\frac{1}{800}$

Percentage	Decimal	Fraction
50%	0.5	$\frac{1}{2}$
33$\frac{1}{3}$%	0.3^{r}	$\frac{1}{3}$
66$\frac{2}{3}$%	0.6^{r}	$\frac{2}{3}$
25%	0.25	$\frac{1}{4}$
75%	0.75	$\frac{3}{4}$
20%	0.2	$\frac{1}{5}$
40%	0.4	$\frac{2}{5}$
60%	0.6	$\frac{3}{5}$
80%	0.8	$\frac{4}{5}$
$16.6\%^{r}$	0.16^{r}	$\frac{1}{6}$
12.5%	0.125	$\frac{1}{8}$
37.5%	0.375	$\frac{3}{8}$
62.5%	0.625	$\frac{5}{8}$
87.5%	0.875	$\frac{7}{8}$
$11.1\%^{r}$	0.11^{r}	$\frac{1}{9}$
$22.2\%^{r}$	0.22^{r}	$\frac{2}{9}$

Percentage converter drill

Set a stopwatch and aim to complete each of the following 10 question drills in 75 seconds.

	Convert to percentage	Convert to decimal	Convert to fraction
Q1	0.25	$^1/_{100}$	87.5%
Q2	0.75	$^1/_{25}$	0.5%
Q3	0.1	$^5/_6$	0.125%
Q4	0.0047	$^3/_8$	60%
Q5	0.063	$^3/_9$	250%
Q6	$^1/_2$	0.1%	0.625
Q7	$^7/_8$	0.043%	2.75
Q8	$^5/_3$	8.3%	0.006
Q9	$^1/_{200}$	430%	0.125
Q10	$^4/_{400}$	$^5/_3$%	6.002

Working with percentages

Many word problems involving percentages require you to work out the missing number in a three-piece puzzle. In percentage problems, the three pieces are the *whole*, the *part* and the *percentage*. The relationship between the three pieces is expressed as:

Part = Percentage × Whole

This can be rearranged as:

$$\text{Whole} = \frac{\text{Part}}{\text{Percentage}}$$

Or in another arrangement:

$$\text{Percentage} = \frac{\text{Part}}{\text{Whole}}$$

When you are working out the answers to the problems, there is often more than one way to arrive at the correct answer. Decide which method is the easiest for you to work with. You can check your answer with an alternative method.

Working with percentages: the part

To find the part when working with a percentage, use the following formula:

Part = Percentage × Whole

Worked example

What is 25% of 25? You are looking for the part of the whole number (25) that equals 25%.

Method 1

Multiply the whole number by the decimal equivalent of the percentage:

25% = 0.25

Percentage	×	Whole	=	Part
0.25	×	25	=	6.25

Method 2

Express the percentage as a fraction over 100 and multiply by the whole number.

$$25\% = \frac{25}{100}$$

$$\frac{25}{100} \times 25 = 6.25$$

Tip: To find 10% of a number, move the decimal point at the end of the number one place to the left. For example, to find 10% of 57.5 move the decimal point 1 place to the left = 5.75.

Working with percentages: the whole

To find the whole when working with a percentage, use the following formula:

$$\text{Whole} = \frac{\text{Part}}{\text{Percentage}}$$

Worked example

25 is 20% of what number? You are looking for a whole number, of which 20% equals 25.

Method 1

When you are working with simple percentages, work out what fraction of the whole quantity the percentage represents. Then calculate by how much you would need to multiply that quantity to obtain the whole.

$$20\% = \frac{1}{5}$$

Therefore:

$$100\% = 5 \times \frac{1}{5}$$

You know that $\frac{1}{5} = 25$, so multiply $\frac{1}{5} \times 5$ to find 100%.
$25 \times 5 = 125$.

Method 2
Set the percentage up as a fraction over 100 and cross-multiply the fractions to find the whole number (x):

$$\frac{20}{100} = \frac{25}{x}$$

Cross-multiply each side:

$20x = 2500$

Divide both sides by 20:

$x = 2500 \div 20$
$x = 125$

Working with percentages: the percentage

To find the whole when working with a percentage, use the following formula:

$$\text{Percentage} = \frac{\text{Part}}{\text{Whole}}$$

Worked example
What percentage of 240 is 30? You are given the part (30) and the whole number (240) and asked to work out the percentage.

Method 1
Use the formula and substitute actual values:

$$x\% = \frac{\text{Part}}{\text{Whole}} \qquad x\% = \frac{30}{240}$$

Reduce the fraction to its lowest terms:

$$x\% = \frac{30}{240} = \frac{1}{8}$$

Convert the fraction to a percentage:

$$x\% = \frac{1}{8} \textit{ or } 12.5\%$$

Method 2

Set up the part and the whole number as a fraction and cross-multiply the fractions to find the percentage (x):

$$\frac{30}{240} = \frac{x}{100}$$

Cross multiply each side of the fraction:

$$240x = 3000$$

Divide each side by 240:

$$x = \frac{3000}{240}$$

$$x = 12.5\%$$

Percentage formulae practice questions

Find the part

Q1 What is 50% of 36?
Q2 What is 75% of 160?
Q3 What is 30% of 130?
Q4 What is 24% of 24?
Q5 What is 20% of 12.5% of 80?

Find the whole

Q6 15 is 25% of what number?
Q7 125 is 62.5% of what number?
Q8 32% of 60 is 60% of what number?
Q9 25 is $16\frac{2}{3}$% of what number?
Q10 112 is 87.5% of what number?

Find the percentage

Q11 18 is what percentage of 48?
Q12 What percentage of 160 is 4?
Q13 12 is what percentage of 200?
Q14 What is 37.5% of 12.5% of 512?
Q15 2.4 is what percentage of 320?

Increasing or decreasing a value by a percentage

Percentage increases and decreases are one of the most common calculations you will be asked to perform in aptitude tests. Being familiar with basic formulae and estimating techniques will save you a lot of time during the test. The key to working out a percentage change quickly and correctly is to estimate the answer first. This will lead you to the right order of operations and will help you to identify the place where you made an error in your calculation, should you find that your original estimate and calculation are way off.

There are two basic methods to work out a percentage increase or decrease. Pick the method that suits you (or your numbers) and use the other to validate your answer.

Method 1

Find the actual value of the percentage increase (or decrease) and add the value of the increase (or decrease) to the original value.

Percentage increase formula

$$\% \text{ increase} = \frac{\textit{Actual amount of increase}}{\textit{Original whole}} \times 100\%$$

New value = Original whole + Amount of increase

Percentage decrease formula

$$\% \text{ decrease} = \frac{\textit{Actual amount of decrease}}{\textit{Original whole}} \times 100\%$$

New value = Original whole – Amount of decrease

Worked example, percentage increase method 1

The price of a barrel of oil increased from \$20 to \$24 between January and April 1992. By what percentage did the price of oil increase during this period?

Estimate the answer

An increase of 50% on the original whole amount of \$20 would give a new price of \$30 (\$20 + (0.5 × \$20) = \$30). An increase of 10% would give a new price of \$22. (\$20 + (0.1 × \$20) = \$22). So you know that your answer will be in the range of 10–50% and will be closer to 10% (\$22) than to 50% (\$30).

Calculate the answer

Use the formula to find a percentage increase and plug in the numbers.

$$\% \text{ increase} = \frac{\textit{Actual amount of increase}}{\textit{Original whole}} \times 100\%$$

The actual amount of increase = the difference in price between January and April = \$24 – \$20 = \$4:

$$\% \; increase = \frac{\$4}{\$20} \times 100\%$$

Reduce the fraction to its lowest terms:

$$\% \; increase = \frac{1}{5} \times 100\%$$

Percentage increase = 20%

Worked example, percentage decrease method 1

The travel agent list price of a flight from London to Santa Domingo is £600. If you purchase the flight online direct from the airline, you can buy the flight for a one-off discounted price of £510. By what percentage is the flight discounted if you buy online?

Estimate the answer

The cost of a £600 flight after a 10% decrease is £600 – £60 = £540. After a 20% decrease, the same flight would cost £600 – (2 × £60) = £480. So you know that your answer will be somewhere between 10% and 20%.

Calculate the answer

Use the formula to find a percentage decrease and plug in the numbers.

$$\% \text{ decrease} = \frac{\textit{Actual amount of decrease}}{\textit{Original whole}} \times 100\%$$

The actual amount of decrease = original whole (£600) – new whole (£510) = £90.

$$\% \textit{ decrease} = \frac{90}{600} \times 100\%$$

Reduce the fraction to its lowest terms:

$$\frac{90}{600} = \frac{3}{20}$$

Percentage decrease = 15%

Method 2

To find the new value multiply by the original whole + percentage increase:

= 1 + percentage increase (expressed as a decimal)

To find the new value multiply by the original whole – percentage decrease:

= 1 – percentage decrease (expressed as a decimal)

Worked example, percentage increase method 2

The wholesale price of a bicycle lock before VAT is £8.50. For how much, to the nearest penny, does the lock sell retail? VAT in the UK is currently 17.5%.

Estimate the answer

A 10% tax on an item costing £8.50 is £0.85. A 20% tax would be (2 × 10%) = £1.70. A 17.5% tax would increase the original whole by an amount between 85p and £1.70. The answer will be between (£8.50 + £0.85) and (£8.50 + £1.70).

Calculate the answer

To find the new value multiply by the original whole + percentage increase:

Original whole	×	(1 + percentage increase expressed as a decimal)	
£8.50	×	1 + 0.175	=
£8.50	×	1.175	= £9.99 (to nearest penny)

Worked example, percentage decrease method 2

A wholesale tyre company offers a discount of 8% to its preferred customers. If a set of four tyres sells for £160 to regular customers, how much does the set cost to a preferred customer?

Estimate the answer

A 10% discount would reduce the price by £16, so your approximate answer will be £160 – £16 = £144.

Calculate the answer

To find the new value multiply by the original whole – percentage decrease. Calculate the decrease by subtracting the amount of percentage change from 100%. In this question, the percentage decrease is 8%, so 1 – 0.08 = 0.92:

Original whole	×	(1 – percentage decrease expressed as a decimal)	
£160	×	1 – 0.08	=
£160	×	0.92	= £147.20

Combining percentages

Percentage questions will often require that you complete the question in several stages. The questions are not difficult as long as you remember a couple of basic rules. First, remember that you cannot just combine all the percentages in a question and find the total percentage of the original whole. Second, always identify the original whole as the starting point.

Worked example

The average price of a London flat in 1970 was £40,000. By the end of the 1980s, the price had risen by 85%. By January 1992, house prices fell dramatically and the price of the same London flat, now beset with structural problems, fell by 40% on the December 1989 price. At how much more or less is the flat valued in 1992 than in 1970?

You are asked to find a new value following a percentage increase and then a percentage decrease. Increase the original whole (£40,000) by 85% (1 + 0.85) to find the December 1989 value:

40,000 × 1.85 = £74,000
New whole = £74,000

Decrease the new whole (£74,000) by 40% (1 – 0.4) to find the January 1992 value:

£74,000 × (1 – 0.4) =
£74,000 × 0.6 = £44,400

The flat is worth £4,400 more in 1992 than the 1970 value.

Percentage increase and decrease practice questions

Q1 Increase 25% of 18 by 15% of 60.

Q2 What is the percentage profit a secondhand car dealer makes on a car he buys for £2,400 and sells for £3,200?

Q3 The share price for stock S rose by 20% on the first day of trading. In the first few minutes of trading the next day, the stock rose by a further 2.5%. By what percentage has the stock gained in price?

Q4 An apple tree grows by 5% every year. When it is planted, its height is 90 cm. In which year will the tree exceed 1m in height?

Q5 Between 2000 and 2001, Kinji's salary increased from £25,000 to £26,500. His projected salary increase for 2002 is 0.5% higher than the increase he received in 2001. What is Kinji's projected salary for 2002?

Q6 If P is discounted by 25%, the new price of P is 12.5% of Q. What is P in terms of Q?

Q7 A child's bike goes on sale at a 15% discount at the end of the summer holidays. By the October half-term holiday, the bike is marked down by a further 20% on the discounted price. If the original price of the bike was £125, what is the selling price of the bike after the October markdown?

Q8 A washing machine goes on sale for £289, which is a 66⅔% markdown on the original price. What was the original price?

Q9 A publisher sells a puzzle book to wholesalers for £6 for the first 21 books purchased and applies a 4% discount to each book purchased thereafter. If the wholesaler buys 30 puzzle books from the publisher, how much does the wholesaler save on the retail price of the order, if the retail price of the book is £7.99?

Q10 The distance between Lovetts Bay and Noe Valley is 124 miles and a car uses 25 litres of petrol to make the journey. After the engine has been tuned, fuel consumption is reduced by 2%. How much petrol is required for a round trip?

Simple interest and compound interest

Simple interest is the amount of interest earned on an investment. *Compound interest* is the amount of interest earned on an investment plus interest earned on previously earned interest. A question that requires you to calculate interest is designed to test your ability to decide whether to perform a simple percentage change calculation or to combine cumulative percentage changes. The clue will usually be hidden in the question, so take your cue from the question itself, before performing any calculations.

Simple interest

To find simple interest, multiply the *principal sum* (usually the original amount invested) by the *interest rate* by the *time period* (usually expressed in years):

Interest = Principal sum × Interest rate × Time period

Worked example

If Alex takes out a two-year loan of £2,000 from a bank, and pays simple annual interest of 8.75%, how much interest will he have paid by the end of the term? To work out how much interest is paid in one year, multiply £2000 (the principal sum) by the decimal equivalent of 8.75%:

£2000 × 0.0875 = £175

In one year, £175 will be paid in interest. In two years, £175 × 2 will be paid (£350). Use the formula and plug in the numbers to verify your answer:

Principal sum	×	Interest rate	×	Time period	=	Interest
£2000	×	0.0875	×	2	=	£350

If the question asked you what is the total amount in the account at the end of the period, you would multiply by (1 + 0.0875). This is very important in compound interest calculations. For example:

Principal sum	×	(1 + Interest rate)	×	Time period	=	Interest plus Principal sum
£2000	×	(1 + 0.0875)	×	2	=	£2,350

Compound interest

To find the compound interest payable on a principal sum, work out the interest on the principal sum for every year interest is accumulated. In compound interest, the charge is calculated on the sum loaned plus any interest accrued in previous periods.

The formula to find compound interest is $I = P\ (1 + R)^{n-1}$ where P = the principal sum, R = the rate of interest and n = the number of periods for which interest is calculated.

Worked example

Shifty borrows £500 over 2 years from a building society at a rate of 12% per annum compounded quarterly. How much interest will Shifty have to pay at the end of the 2-year loan?

Recall the formula to find compound interest: $I = P(1 + R)^{n-1}$.

If £500 is loaned for 2 years at a rate of 12% per annum, compounded quarterly, the calculations need to be made on a quarterly basis. So the value of n will be 4 (quarters) × 2 (years) = 8, and the value of r will be $^{12}\!/_{4}$ = 3% (per quarter). Therefore $I = 500^{\ (1.03)8-1} = £\ 133.38$.

Simple and compound interest practice questions

Q1 Jo invests £9,750 in an online savings account for 1 year and 6 months. Simple interest is paid at a rate of 3.75% per year and is calculated on a daily basis. To the nearest penny, how much is in the account at the end of the term?

Q2 Justin has £2,100 in his savings account. He deposited £1,500 two years ago. What is the simple annual interest rate on the account?

Q3 Katie invested £60 in a bond which yields a simple annual interest rate of 1.25%. If the total amount payable to Katie at the end of the term is £63.75, after how many years has Katie redeemed the bond?

Q4 Max bought a house for £40,000. The value of his house increased over 6 years, so that by the end of the term it had seen a 40% increase on the original price. The same amount invested in a bank account yields 9% per annum simple interest over the same period. Which is the more profitable investment option?

Q5 If Paul invests £385 in National Savings Certificates, which currently yield annual compound interest of 2.8%, what will his certificates be worth at the end of 3 years (to the nearest penny?)

Answers to Chapter 4

Percentage converter drill

	Convert to percentage	Convert to decimal	Convert to fraction
Q1	$0.25 = 25\%$	$\frac{1}{100} = 0.01$	$87.5\% = \frac{7}{8}$
Q2	$0.75 = 75\%$	$\frac{1}{25} = 0.04$	$0.5\% = \frac{1}{200}$
Q3	$0.1 = 10\%$	$\frac{5}{6} = 0.83^{r}$	$0.125\% = \frac{1}{800}$
Q4	$0.0047 = 0.47\%$	$\frac{3}{8} = 0.375$	$60\% = \frac{3}{5}$
Q5	$0.063 = 6.3\%$	$\frac{3}{9} = 0.3^{r}$	$250\% = 2\frac{1}{2}$
Q6	$\frac{1}{2} = 50\%$	$0.1\% = 0.001$	$0.625 = \frac{5}{8}$
Q7	$\frac{7}{8} = 87.5\%$	$0.043\% = 0.00043$	$2.75 = 2\frac{3}{4}$
Q8	$\frac{5}{3} = 166.6\%^{r}$	$8.3\% = 0.083$	$0.006 = \frac{6}{1000}$
Q9	$\frac{1}{200} = \frac{1}{2}\%$	$430\% = 4.3$	$0.125 = \frac{1}{8}$
Q10	$\frac{4}{400} = 1\%$	$\frac{5}{3}\% = 0.16^{r}$	$6.002 = 6\frac{1}{500}$

Percentage formulae practice questions: explanations

In the following explanations, only one method is given. You may prefer to use another method to check your answers.

Q1 Answer = 18

Convert 50% to its decimal equivalent and multiply the whole number by the decimal:

$50\% = 0.5$

$0.5 \times 36 = 18$

Q2 Answer = 120

Convert 75% to its decimal equivalent and multiply the whole number by the decimal:

75% = 0.75
0.75 × 160 = 120

Q3 Answer = 39

To find 10% of 130, move the decimal place one place to the left of the whole number:

10% = 13
30% = 3 × 10% = 39

Q4 Answer = 5.76

24% is approximately equal to ¼ (25%) so your answer will be approximately ¼ × 24 = 6. Express the percentage as a fraction over 100 and multiply by the whole number (24):

$$\frac{24}{100} \times 24 = 5.76$$

Q5 Answer = 2

Both percentages are recognizable as common fractions, so express both as fractions and multiply by the whole number (80).(You can also find 20% of 12.5% (= 2.5%) and multiply by the whole number, 80.)

$$20\% = \frac{1}{5} \text{ and } 12.5\% = \frac{1}{8}$$

$$\frac{1}{8} \times 80 = 10$$

$$\frac{1}{5} \times 10 = 2$$

Q6 Answer = 60

25% = ¼, so you know that 15 = ¼ x (where x = the whole number). To solve the equation for a whole number, multiply both sides by the inverse of the fraction.

$$\frac{4}{1} \times 15 = \frac{4}{1} \times \frac{1}{4} x$$

$$60 = x$$

Q7 Answer = 200

$$62.5\% = \frac{5}{8} \text{ so } 125 = \frac{5}{8} x$$

To solve the equation for a whole number, multiply both sides by the inverse of the fraction:

$$\frac{8}{5} \times 125 = \frac{8}{5} \times \frac{5}{8} x$$

Reduce the equation to its simplest terms:

$$\frac{8}{1} \times 25 = x$$

$$200 = x$$

Q8 Answer = 32

Convert the percentage to its decimal equivalent: 32% = 0.32. Multiply by the whole number (60) to find the percentage to find the part:

Percentage	×	**Whole**	=	**Part**
0.32	×	60	=	19.2

So you know that 19.2 = 60% of a new whole number x. Solve for x by setting up the percentage as a fraction over 100 and cross-multiply:

$$\frac{(part)}{(whole)} \quad \frac{19.2}{x} = \frac{60}{100}$$

$60x = 19.2 \times 100 = 1920$. Divide both sides by 60:

$$x = \frac{1920}{60}$$

$$x = 32$$

Q9 Answer = 150

Recall $16\frac{2}{3}\% = \frac{1}{6}$

So $25 = \frac{1}{6}x$

To solve the equation for a whole number, multiply both sides by the inverse of the fraction:

$$\frac{6}{1} \times 25 = \frac{6}{1} \times \frac{1}{6}x$$

$$150 = x$$

Q10 Answer = 128

Recall $87.5\% = \frac{7}{8}$

So $112 = \frac{7}{8}x$

To solve the equation for a whole number, multiply both sides by the inverse of the fraction:

$$\frac{8}{7} \times 112 = \frac{8}{7} \times \frac{7}{8}x$$

Reduce the equation to its simplest terms:

$$\frac{8}{1} \times 16 = x$$

$$128 = x$$

Q11 Answer = 37.5%

$$\frac{\text{Part}}{\text{Whole}} = \text{Percentage}$$

Plug in the numbers and solve for x:

$$\frac{(part)\ \ 18}{(whole)\ 48} = x\%$$

$$\frac{3}{8} = x\%$$

Recall from the converter table that $\frac{3}{8} = 37.5\%$.

Q12 Answer = 2.5%

$$\frac{\text{Part}}{\text{Whole}} = \text{Percentage}$$

Plug in the numbers to the formula:

Reduce the fraction to its lowest terms:

$$\frac{(part)}{(whole)} \; \frac{4}{160} = Percentage$$

You know from the converter table that $\frac{1}{20}$ = 5% and therefore $\frac{1}{40}$ = 2.5%.

$$\frac{1}{40} = Percentage$$

Q13 Answer = 6%

12 ÷ 200 = 0.06 = 6%

$$\frac{(part)}{(whole)} \; \frac{12}{200} = percentage$$

Q14 Answer = 24

First find 12.5% of the whole number (512):

$$12.5\% = \frac{1}{8} \; and \; 37.5\% = \frac{3}{8}$$

$$\frac{1}{8} \times 512 = 64$$

Next find 37.5% of the new whole number:

$$\frac{3}{8} \times 64 = 24$$

Q15 Answer = 0.75%.

Percentage	×	Whole	=	Part
x %	×	320	=	2.4

Divide both sides by 320:

$$x\% = \frac{2.4}{320} = 0.75\%$$

Percentage increase and decrease practice questions

Q1	13.5
Q2	33⅓%
Q3	23%
Q4	Year 3
Q5	£28,222.50
Q6	P = $^{9}/_{6}$
Q7	£85
Q8	£867
Q9	£61.86
Q10	49 litres

Percentage increase and decrease practice questions explanations

Q1 Answer = 13.5

First find 25% of 18:

$$\frac{25}{100} \times 18 = \frac{1}{4} \times 18 = 4.5$$

Now find 15% of 60:

$$\frac{15}{100} \times 60 = \frac{3}{20} \times 6 = 9$$

The question asks you to increase 4.5 by 9 = 4.5 + 9 = 13.5

Q2 Answer = 33⅓%

Use the percentage increase formula:

$$\%\ increase = \frac{Actual\ amount\ of\ increase}{Original\ whole} \times 100\%$$

New whole – original whole = actual amount of increase:

3200 – 2400 = 800

Plug in the numbers to the formula:

$$\%\ increase = \frac{800}{2400} \times 100\%$$

Reduce the fraction to its lowest terms:

$$\frac{1}{3} = 33\frac{1}{3}\%$$

Q3 Answer = 23%

Let S represent the price of the stock.
Price of stock S following a 20% increase = S (1.2).
Price of stock S following a 20% and 2.5% increase = S(1.2) (1.025)
Final price = S(1.23).
1.23 represents a 23% increase.

Q4 Answer = Year 3

This question requires you to work out the height of the tree each year as a percentage increase on the height of the previous year. To increase by 5%, multiply the whole by 1.05:

Year	Whole	×	5% increase	=	New whole
1	90 cm	×	1.05	=	94.5 cm
2	94.5 cm	×	1.05	=	99.225 cm
3	99.225 cm	×	1.05	=	104.18625 cm

Q5 Answer = £28,222.50

Work out the percentage increase in salary between 2000 and 2001:

$$\%\ increase = \frac{Actual\ amount\ of\ increase}{Original\ whole} \times 100\%$$

New whole – original whole = actual amount of increase:

£26,500 – £25,000 = £1,500

Plug in the numbers to the formula:

$$\%\ increase = \frac{1500}{25000} \times 100\%$$

Reduce the fraction to its lowest terms:

$$\%\ increase = \frac{3}{1} \times 2\%$$

Percentage increase = 6%. Projected 2002 salary = £26,500 (6% + 0.5%)
£26,500 × (1 + 0.065) = £28,222.50

Q6 Answer $P = \frac{Q}{6}$

Pick numbers for this question and work through the question. Let P = 100.
First discount P by 25%:

100 × (1 – 0.25) = 75

The new price of P (=75) = 12.5% of Q:

$$(75) = \frac{1}{8}Q$$

Multiply by the reciprocal of the fraction to solve for Q:

$$\frac{8}{1}(75) = \frac{8}{1} \times \frac{1}{8}Q$$

600 = Q and P = 100 and therefore:

$$P = \frac{Q}{6}$$

Q7 Answer = £85
A 15% discount is the same as saying that the bike will cost 85% of the original value (100% –15% = 85%). A further 20% discount is the same as 80% of the new value.
First discount the original price by 15%:

£125 × (1 – 0.15) = £106.25

Now discount the new whole (= £106.25) by a further 20%:

£106.25 × (1 – 0.2) = £85

Q8 Answer = £867

Recall that $66\frac{2}{3}\%$ = 2/3.
Sale price = 289; Original price = O. Now solve for O:

$$O - \left(\frac{2}{3}O\right) = 289$$

Now multiply both sides by the reciprocal of ⅔ × O:

$$\frac{3}{2}O - \frac{3}{2} \times \frac{2}{3}O = \frac{3}{2} \times 289$$

Now simplify the equation:

$$\frac{3}{2}O - O = 433.5$$

Multiply both sides by 2:

O = 2 × 433.5 = 867

Alternatively, use logic to approximate the answer. If 289 = 33.3%, then 3 × 289 = 100%.

Q9 Answer = £61.86

The first 21 books cost £6 each = 21 × £6 = £126.
The next 9 books are discounted by 4% on the wholesale price of £6:

Number of books	×	**(Original price 4% discount)**	=	**Discounted price**
9 books	×	(£6 × (1 – 0.04))	=	£51.84

Total cost of wholesale order = (£126 + £51.84) = £177.84
The retail price of 30 books = £7.99 × 30 = £239.70
Retail price – Wholesale price = Amount saved
£239.70 – £177.84 = £61.86

Q10 Answer = 49 litres

The round trip of 248 miles requires (2 × 25 litres) = 50 litres. After the engine tune, fuel consumption is reduced by 2%. Decrease 50 litres by 2%:

50 × (1 – 0.02) = 49 litres

Simple and compound interest practice questions

Q1 £10,298.44
Q2 20%
Q3 5 years
Q4 Bank account
Q5 £418.25

Simple and compound interest practice questions explanations

Q1 Answer = £10,298.44

Use the formula and plug in numbers:

Principal sum	×	**Interest rate**	×	**Time period**	=	**Interest**
£9750	×	3.75%	×	1½ years	=	Interest
£9750	×	0.0375	×	1.5	=	£548.44 (to nearest p)

Total interest paid = £548.44. The question asks for the total amount in the account at the end of the term, so add the interest to the principal sum: £9750 + 548.44 = £10,298.44.

Q2 Answer = 20%

First work out the actual amount of interest earned on the account:

Final sum –	Principal sum =	Total interest
£2100 –	£1500 =	£600

Total time = 2 years. Plug in the numbers to the formula:

$$\textit{Simple annual interest} = \frac{600}{2} = 300$$

$$\textit{Interest} = \frac{(\textit{Interest})}{(\textit{Principal sum})}\ \frac{300}{1500} \times 100 = 20\%$$

Q3 Answer = 5 years

After 1 year, simple interest earned =
Principal sum × Interest (£60 × 1.25%)
£60 × 0.0125 = £0.75 per year simple interest.

Total interest earned during the term =
(Principal sum + interest) – Principal sum
£63.75 – £60 = £3.75

Divide the annual interest (£0.75) by the total interest (£3.75) to find the time period:

$$\frac{3.75}{0.75} = \textit{5 years}$$

Q4 Answer = bank account

Value of the house after 6 years = Principal sum × 40% increase
£40,000 × 1.4 = £56,000

Value of investment in the bank account =
Principal sum + (Principal sum × Interest × Time period)
£40,000 + (£40,000 × 0.09 × 6 years) = £61,600

The bank account investment yields an additional £5,600.

Q5 Answer = £418.25

You can either use the formula to work out the amount of compound interest, or work out the total value of the certificate at the end of the period by working out the value of the certificate year on year.

Year	Principal sum	×	Interest (1 + 2.8%)	=	Principal sum + Interest
1	£385	×	1.028	=	£395.78
2	£395.78	×	1.028	=	£406.86
3	£406.86	×	1.028	=	£418.25

5

Ratios and proportions

Chapter topics

- Terms used in this chapter
- Working with ratios
- Ratios and common units of measure
- Types of ratio
- Using ratios to find actual quantities
- Proportions
- Answers to Chapter 5

Terms used in this chapter

Proportion: Equality of ratios between two pairs of quantities.
Ratio: The comparison between two or more quantities.

When you define the relationship between two (or more) quantities of the same kind you are finding a *ratio*. A ratio tells you the relationship between quantities, but does not necessarily tell you the actual quantities. For example, the instruction 'to bake a scone, use half the quantity of fat to one quantity of flour' tells you the ratio of fat to flour, but does not tell you how much of each quantity to use. When you compare two ratios, you are finding a *proportion*. In this chapter you will learn first how to work with ratios, and then you will apply that knowledge to understand proportions and when to use them.

Working with ratios

Ratios work in a similar way to fractions and are usually reduced to the lowest term. In aptitude tests, ratios are typically used to represent relative quantities of whole units, such as the ratio of adults to children in a playground, or the ratio of inches to miles on a map, or of gin to vermouth in a martini. You can set up a ratio in several different ways. For example, to define the relationship between 5 black cats and 10 white cats:

Method 1

Define the ratio in words, separating the quantities with 'to'. For example:

The ratio of black cats to white cats is five 'to' ten.

Method 2

Separate the quantities with a colon, where the colon replaces the word 'to'. For example:

The ratio of black cats to white cats is 5 : 10.

This can be reduced to its lowest terms like an equation. Simply divide both sides by the same number, so 5 : 10 = 1 : 2 when you divide both sides by 5.

Method 3

Write the ratio as a fraction in its lowest terms. For example:

$$\frac{\textit{Black cats}}{\textit{White cats}} = \frac{5}{10} = \frac{1}{2}$$

A simple way to remember which way to set up the fraction is with a formula:

$$\textit{Ratio} = \frac{\text{'of' } x}{\text{'to' } y}$$

For example, what is the ratio of black cats to white cats?

$$\text{Ratio} = \frac{\text{'of' } \textit{black cats}}{\text{'to' } \textit{white cats}} = \frac{5}{10}$$

Reduce the fraction to its lowest terms.

$$\frac{5}{10} = \frac{1}{2}$$

Worked example

Express the following ratios (1) in words (2) with a colon separating the quantities and (3) as a fraction.

Q1 What is the ratio of shoes to feet?

(1) Shoes to Feet

(2) Shoes : Feet

(3) $\frac{\text{Shoes}}{\text{Feet}}$

Q2 What is the ratio of flowers to stems?

(1) Flowers to Stems

(2) Flowers : Stems

(3) $\frac{\text{Flowers}}{\text{Stems}}$

Q3 What is the ratio of guitarists to bandmembers?

(1) Guitarists to Bandmembers

(2) Guitarists : Bandmembers

(3) $\frac{\text{Guitarists}}{\text{Bandmembers}}$

Ratios and common units of measure

In order to compare two quantities, they must be expressed in terms of the same unit. To set up a ratio when the original units of measure are in different units, simplify the calculation by converting the larger unit to the smaller unit of measure. This method ensures that you work with common units throughout your calculation.

Worked example

What is the ratio of 20 minutes to 2 hours?

Convert the larger unit (2 hours) into the smaller unit (minutes): 2 hours = 120 minutes.

The ratio of 20 minutes to 120 minutes = 20 : 120.

Divide both sides by 20:

Ratio = 1 : 6.

Worked example

What is the ratio of 25 grammes (g) to 5 kilogrammes (kg)? Convert the larger unit (kg) to the equivalent measure of the smaller unit (g): 1 kg = 1,000 g so 5 kg = (5 × 1,000 g) = 5,000 g

Therefore the ratio of 25 g : 5 kg = 25 : 5,000

Divide both sides by 25 to express the ratio in its lowest terms.

The ratio of 25 g : 5 kg in its lowest terms is 1 : 200.

Ratios converter drill 1

Set a stopwatch and aim to complete the following drill in two minutes. Express the following ratios in their simplest forms, in the format x : y.

Q1 $\frac{1}{4}$ Q2 $\frac{6}{5}$

Q3 $\frac{20}{5}$ Q4 $\frac{87}{87}$

Q5 $\frac{12}{15}$ Q6 $\frac{360}{60}$

Q7 $\frac{21}{9}$ Q8 $\frac{138}{144}$

Q9 $\frac{105}{135}$ Q10 $\frac{115}{85}$

Ratios converter drill 2

Set a stopwatch and aim to complete the following drill in 1 minute. Express the following ratios, given in the format x : y, as fractions in their lowest form.

Q1	22 : 3
Q2	19 : 20
Q3	3 : 15
Q4	35 : 45
Q5	3 : 4
Q6	325 : 375
Q7	105 : 120
Q8	7 : 22
Q9	2 : 56
Q10	268 : 335

Types of ratio

Two types of ratio are typically tested in numeracy tests. These are 'part to part' ratios and 'part to whole' ratios. When you think about the 'whole' and the 'part' with ratios, think of the 'whole' as the complete set (or the 'parent set') and the 'part' as a subset of the parent set. For example, players in a football squad are the complete set (the whole) and those players picked for the starting line-up are the subset (the part). Once you understand how these ratios work, you can use ratios to determine actual values, for example the exact number of players in the squad.

'Part to part' ratios

Worked example

There are four oak trees for every two willow trees. What is the ratio of willows to oaks?

'Types of trees' is the parent set, and consists of two subsets, 'willows' and 'oaks'.

Willows to Oaks = 2 : 4

Divide both sides by 2 to express the ratio in its simplest form:

Willows to Oaks = 1 : 2

'Part to whole' ratios

If you know the ratio of subsets to the complete set, you can work out the number of parts in the whole. You can also convert the 'part to part' ratio to a 'part to whole' ratio only when you are sure that there are no missing parts.

Worked example

A filing cabinet in Whitehall contains only files classified as 'secret' or 'top secret'. The ratio of 'secret' to 'top secret' files in the filing cabinet is 3 : 2. What is the ratio of 'top secret' files to all files in the cabinet?

The parent set is 'all files in the filing cabinet' and consists of two subsets, 'secret files' and 'top secret files'. You are given a 'part to part' ratio and asked to express one part as a subset of the whole. As there are only 'secret' and 'top secret' files in the filing cabinet, you can convert the 'part to part' ratio to a 'part to whole' ratio:

Part (secret) +	Part (top secret) =	Whole
3 +	2 =	5

This tells you that for every five files, three are classified as 'secret' and two are classified as 'top secret'. The ratio of top secret files to all files is therefore:

$$\frac{Part}{Whole} = \frac{2}{5} \quad or\ 2:5$$

Note that this does not give you actual values (the number of files), just the ratio between quantities.

Worked example

A software company decides to expand its floor area by building additional floors underground for the software testers. When the construction is finished, one in four of the total floors in the building will be underground. What is the ratio of floors above ground to floors underground?

The floors can only be underground or above ground, so you can be sure to know all the subsets of the parent set. Therefore you can determine the part to whole ratio. The question tells you that for every floor underground, three are above ground. If ¼ of all floors are below ground, then ¾ must be above ground. The ratio of floors above ground to floors below is ¾ to ¼. Both quantities are expressed as quarters, so you can form a ratio with the numerators:

Floors above ground to floors below ground = 3 : 1.

Using ratios to find actual quantities

A frequently tested concept is the use of ratios to find actual quantities. Ratios only tell you the relationship between numbers, not actual quantities. However, if you know the actual quantity of either a part or the whole, you can determine the actual quantity of the other parts.

Worked example

In a bag of 60 green and red jellybeans, the ratio of red jellybeans to green jellybeans is 2 : 3. How many are red?

Method 1

The parent set (the whole) is 'all jellybeans in the bag'. The subsets (the parts) are 'red jellybeans' and 'green jellybeans'. You know the actual value of the whole and the ratio between the parts, so you can work out the actual values of each part:

Part (red) +	Part (green) =	Whole
2 +	3 =	5

This tells you that for every 5 parts, 2 are red and 3 are green. To find the actual quantity of red jellybeans, multiply the ratio of red jellybeans by the actual value of the whole.

$$\frac{2}{5}(60) = 24$$

Method 2

You can solve this with simple algebra.
One part = x. Therefore $2x$ (red) + $3x$ (green) = 60 (all jellybeans).
$5x = 60$
$x = 60 \div 5 = 12$
$2x = 24$ red jellybeans.

Worked example

A certain spice mix contains 66 g of a mix of cumin and coriander in the ratio of 4 : x. If 1 part cumin = 6 g, what is the ratio of cumin to coriander?

The parent set is '66 g of spice mix' and consists of two subsets, cumin and coriander.

1 part cumin = 6 g, so 4 × 6 g cumin = 4 parts (24 g). If 24 g of the mix is cumin, then 66 g – 24 g (42 g) is coriander.

The ratio of actual quantities of cumin to coriander = 24 g : 42 g. Divide both sides by 6 to comply with the format of the ratio in the question: 24 : 42 = 4 : 7. The ratio of cumin to coriander is 4 : 7.

Worked example

At the Tedbury Rolling-Rollers, a race open to skaters and bladers, the ratio of skaters to all racers is 2 to 3. There are a total of 240 Rolling-Rollers in the race. How many are on skates?

You are given a 'part to whole' ratio and asked to work out the actual value of a part:

$$\frac{\textit{Part (skaters)}}{\textit{Whole (skaters \& bladers)}} = \frac{2}{3}$$

The total number of competitors is 240. Multiply the ratio (expressed as a fraction) by the actual value of the whole number to find the number of skaters:

$$\frac{2}{3} \times 240 = 160$$

There are 160 skaters in the race.

Worked example

After a hot whites wash, the ratio of the pink socks to all socks emerging from the washing machine is 3 : 5. 15 pairs of socks were put in the washing machine at the start of the programme. How many socks are not pink at the end of the wash? (Assume that all the socks that go into the washing machine also come out of it.)

15 pairs of socks went into the washing machine = 30 socks. The question tells you that the ratio of pink socks to all socks = 3 : 5. Express the ratio as a fraction:

$$\frac{\textit{Part}}{\textit{Whole}} = \frac{3}{5}$$

If $\frac{3}{5}$ of the socks are pink, then $\frac{2}{5}$ of the socks are not pink. Multiply $\frac{2}{5}$ by the actual value of the whole:

$$\frac{2}{5} \times 30 = 12$$

Number of non-pink socks = 12 (or 6 pairs).

Ratios practice questions

Give all your answers in the ratio format x : y.

Q1 In a school there is one qualified teacher for every 32 students. What is the ratio of students to qualified teachers?

Q2 Last season, Southwold United won 18 games and lost 9 games. What is the ratio of games lost to games won?

Q3 It takes 1 hour 20 minutes to bake a potato and 45 minutes to bake a pie. What is the ratio of time taken to bake a potato to time taken to bake a pie?

Q4 Dave's training schedule requires him to increase his weekly mileage by 50 miles every week. In Week 2 he cycles 250 miles. What is the ratio of miles cycled in week 2 to miles cycled in week 3?

Q5 What is the ratio of currants to sultanas in a fruitcake consisting of 1lb 2 oz currants and 12 oz sultanas? (There are 16 oz in 1 lb.)

Q6 In a musical ensemble, the ratio of stringed instruments to all other instruments is 8 : 3. What is the ratio of stringed instruments to all instruments in the ensemble?

Q7 At a rock concert, 250 out of the capacity crowd of 30,000 are left-handed. What fraction of the crowd is right-handed? (Assume that no one in the crowd is ambidextrous!)

Q8 In a certain yoga class of 32 attendees, 8 can stand on their heads unaided. What is the ratio of those who can stand unaided on their heads to those who can't?

Q9 In a restaurant, the ratio of vegetarians to meat-eaters is 3 to 5. Of the vegetarians, ⅔ eat fish. What is the ratio of vegetarians to vegetarian fish-eaters to meat-eaters?

Q10 The ratio of cups to mugs on a table is 1 : 6. If ¼ of the cups and ⅔ of the mugs are filled with tea, and all the other cups and mugs are filled with coffee, what is the ratio of tea to all the drinks on the table?

The following questions require that you use a ratio to calculate actual values.

Q11 On a world atlas, ½ cm represents 250 miles. What is the distance represented by 2¾ cm?

Q12 A company issues dividend payments to two shareholders, Anne and Paula, in the ratio 5 : 4. Anne receives £225. How much does Paula receive?

Q13 Burnford hockey club 'games won' to 'games lost' record last season was 2 : 3. How many games did they play last season if all the games were either won or lost and Burnford won 6 games?

Q14 Charlotte ran a marathon (26.2 miles) in 4 hours and 30 minutes. She ran the first half in $\frac{4}{5}$ of the time it took to run the second half. How long did the first half take?

Q15 In Factory A, the ratio of paper clip production to pencil sharpener production is 400 to 1 and the ratio of pencil sharpener production to stapler production is 3 to 5. What is the ratio of paper clip production to stapler production?

Proportions

When you compare two equivalent ratios of equal value you are finding a proportion. You can think of proportions as ratios reduced to the simplest terms. For example, if you simplify the ratio 30 computers to 48 televisions you are finding a proportion.

When you know how to work with proportions, you can check easily whether two ratios are equal, find missing terms in a ratio, work out the greater of two ratios and work out proportional changes to a ratio. Knowledge of proportions is a good trick to have up your sleeve in an aptitude test. Swift multiplication of the elements in a proportion will help you to verify the answer to a ratio problem quickly.

Worked example: Method 1

In a proportion, the product of the *outer* terms equals the product of the *inner* terms. Does the ratio 24 apples to 36 apples equal the ratio 2 apples to 3 apples?

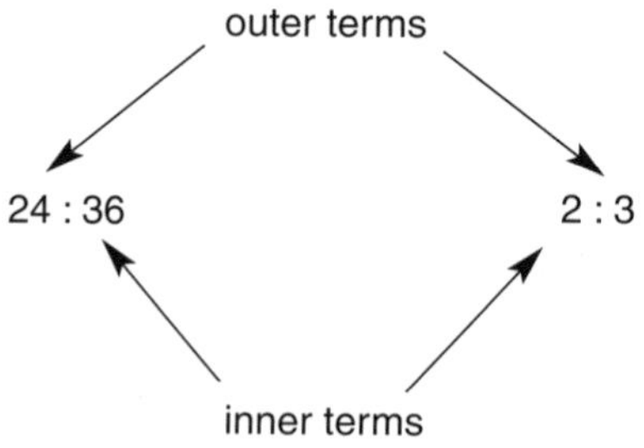

24 × 3	=	36 × 2
72	=	72

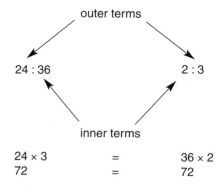

The proportions are equivalent.

Worked example: Method 2

Another way to think about proportions is to set up the ratios as fractions and cross-multiply. Are the following ratios equal: 25 pencils to 5 fountain pens = 100 pencils to 20 fountain pens?

Set up the ratios as fractions.

Tip: It doesn't matter which unit is the numerator, as long as both sides of the equation have the same unit as the numerator.

$$\frac{\textit{Pencils}}{\textit{Fountain pens}} \quad \frac{25}{5} = \frac{100}{20}$$

Cross-multiply the fractions

$$(20 \times 25 = 500) \quad \frac{25}{5} \times \frac{100}{20} \quad (5 \times 100 = 500)$$

The ratios are equal, as both sides when cross-multiplied produce the same result. The ratios therefore form a proportion.

Tip: $a/b = x/y$ is the same as $a : b = x : y$.

Worked example

Is $\frac{2}{3}$ greater than $\frac{8}{13}$? (Is $\frac{2}{3} > \frac{8}{13}$?)

To answer the question, cross-multiply the numerators and denominators:

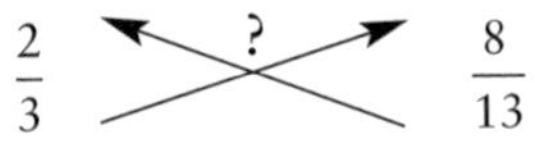

Follow the arrows and write down your answers at the end of the arrowhead.

(2 × 13 = 26)

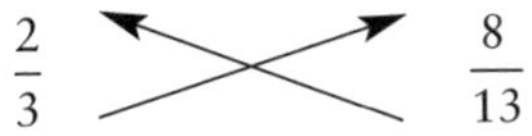

(3 × 8 = 24)

Now ask, 'Is 26 > 24?' The answer is 'yes', so you know that $\frac{2}{3} > \frac{8}{13}$.

Worked example

Use method 2 to find the missing term.

What is the value of x in the following proportion: $2 : 3 = x : 54$?

Set up the ratios as fractions and cross-multiply.

$$\frac{2}{3} \qquad \frac{x}{54}$$

$$\begin{aligned} 3x &= 2 \times 54 \\ 3x &= 108 \\ x &= \frac{108}{3} \\ x &= 36 \end{aligned}$$

Proportions: practice drill 1

Set a stopwatch and aim to complete the following drill in two minutes. Are the following proportions equivalent?

Q1	3 : 7	and	9 : 21
Q2	2 : 5	and	3 : 7
Q3	2 : 5	and	4 : 10
Q4	12 : 92	and	3 : 23
Q5	5 : 27	and	12 : 65
Q6	36 : 54	and	6 : 9
Q7	3 : 19	and	12 : 72
Q8	2 : 7	and	12 : 41
Q9	7 : 12	and	14 : 24
Q10	4 : 124	and	3 : 95

Proportions: practice drill 2

Set a stopwatch and aim to complete the following drill in two minutes. Find the missing terms in the following proportions.

Q1	$\frac{x}{5} \times \frac{8}{20}$	Q2	$\frac{3}{9} \times \frac{162}{x}$
Q3	$\frac{24}{x} \times \frac{6}{3}$	Q4	$\frac{52}{13} \times \frac{4}{x}$
Q5	$\frac{8}{24} \times \frac{x}{30}$	Q6	$\frac{4}{x} \times \frac{16}{20}$
Q7	$\frac{x}{35} \times \frac{7}{5}$	Q8	$\frac{2}{9} \times \frac{6}{x}$
Q9	$\frac{234}{39} \times \frac{468}{x}$	Q10	$\frac{4}{28} \times \frac{3116}{x}$

Proportions practice questions

Solve these questions by setting up a proportion and solving for the missing term.

Q1 In a swimming lesson, there are 4 swimming teachers for every 24 children. How many teachers should there be if there are 36 children in the pool?

Q2 In an evening class, the ratio of students studying Russian to students studying Chinese is 4 : 5. There are 81 students registered to study either Russian or Chinese. How many are studying Chinese?

Q3 A computer image measuring 180 pixels in height and 240 pixels in width is enlarged proportionately. After the enlargement, the picture measures 540 pixels in height. What is its width?

Q4 If four keyboards cost $99, how much do five keyboards cost?

Q5 Find the missing term in the proportion $3 : 9 = x : 18$.

Q6 If 12 doughnuts cost £1.80, how much do 15 doughnuts cost?

Q7 If it takes Emma 10 minutes to do 60 sit-ups, how many minutes (M) does it take her to do S sit-ups?

Q8 Simplify the ratio $16x^2 : 4x^2$.

Q9 The ratio of carers to residents in a residential care home is 1 : 6. If an extension is built to the main building, an additional 40 residents will be accepted. If all 40 places are allocated and the number of carers remains constant, there will be eight times as many residents as carers. How many carers are there currently?

Q10 In a race, for every mountain bike there are 20 road bikes. If an additional 200 road bikers were to join the race, there would be 25 times as many road bikes as mountain bikes in the race. How many road bikes and mountain bikes are there at the beginning of the race?

Answers to Chapter 5

Ratios converter drill 1

Q1 1 : 4
Q2 6 : 5
Q3 4 : 1
Q4 1 : 1
Q5 4 : 5
Q6 6 : 1
Q7 7 : 3
Q8 23 : 24
Q9 7 : 9
Q10 23 : 17

Ratios converter drill 2

Q1 $\frac{22}{3}$

Q2 $\frac{19}{20}$

Q3 $\frac{1}{5}$

Q4 $\frac{7}{9}$

Q5 $\frac{3}{4}$

Q6 $\frac{13}{15}$

Q7 $\frac{7}{8}$

Q8 $\frac{7}{22}$

Q9 $\frac{1}{28}$

Q10 $\frac{4}{5}$

Ratios practice questions

Q1 32 : 1
Q2 1 : 2
Q3 16 : 9
Q4 5 : 6
Q5 3 : 2
Q6 8 : 11
Q7 119 : 120
Q8 1 : 3
Q9 1 : 2 : 5
Q10 17 : 28
Q11 1375 miles
Q12 £180
Q13 15 games
Q14 2 hours
Q15 240 : 1

Ratios practice questions explanations

Q1 Answer = 32 : 1

Recall the formula:

$$\textit{Ratio} = \frac{\text{`}\textit{of}\text{'}}{\text{`}\textit{to}\text{'}}$$

Take your cue from the question and plug in the numbers to the formula:

$$\textit{Ratio} = \frac{\text{`}\textit{of students}\text{'}}{\text{`}\textit{to teachers}\text{'}} = \frac{32}{1}$$

Remember the colon represents 'to' so plug in the answer 32 : 1.

Q2 Answer = 1 : 2

Games Lost = 9
Games Won = 18
Ratio of Lost : Won = 9 : 18
Divide both sides by 9 to reduce the ratio to its simplest form:
Lost to Won = 1 : 2.

Q3 Answer = 16 : 9

Convert the units to the smaller of the two units: 1 hour 20 minutes = 80 minutes.
Potato : Pie = 80 : 45
Divide both sides by 5 to reduce the ratio to its simplest form:
Potato : Pie = 16 : 9.

Q4 Answer = 5 : 6

Week 2 = 250 miles
Week 3 = (250 + 50) miles
Week 2 : Week 3 = 250 : 300
Divide both sides by 50 to reduce the ratio to its simplest terms:
Week 2 : Week 3 = 5 : 6.

Q5 Answer = 3 : 2

Convert the weights to the smaller unit.
Currants = 1 lb 2oz = 18 oz
Currants : Sultanas = 18 : 12
Divide both sides by 6 to reduce the ratio to its simplest form:
Currants : Sultanas = 3 : 2.

Q6 Answer = 8 : 11

You are given a 'part to part' ratio and asked to find a 'part to whole' ratio. 'All instruments' is the parent set and consists of two subsets: 'stringed instruments' and 'non-stringed instruments'.

Part (strings) + Part (non-stringed instruments) = Whole
8 + 3 = 11
Part (strings) to Whole (all instruments) ratio = 8 : 11.

Q7 Answer = 119 : 120

When a question asks you 'what fraction of', you are usually expected to find a part : whole ratio.

If 250 are left-handed, there are (30,000 – 250 = 29,750) right-handed people in the crowd.

$$\frac{\textit{Part (right-handed)}}{\textit{Whole (total in crowd)}} = \frac{29{,}750}{30{,}000}$$

Reduce the fraction to its lowest terms:

$$\frac{29{,}750}{30{,}000} = \frac{119}{120}$$

Now express the ratio in $x : y$ format 119 : 120.

Q8 Answer = 1: 3

You are being asked to find an unknown part in a part : part ratio. You know the actual values of the 'part : whole' ratio: Part (headstanders) to Whole (total in class) = 8 : 32.

Therefore Part (non-headstanders) to Whole (total in class) = (32 – 8) : 32

Headstanders = 8 and Non-headstanders = 24, therefore Headstanders : Non-headstanders = 8 : 24.

Divide each part by 8 to express the ratio in its simplest terms = 1 : 3.

Q9 Answer = 1: 2 : 5

Pick numbers to help you to solve this problem. You are given a part to part ratio of 3 : 5. From this you know that the 'whole' in the 'part to whole' ratio is (3 + 5) = 8. Pick a multiple of 8 to solve the problem, for example 24. So there are 24 diners in the restaurant and each part of the ratio represents (24 ÷ 8) = 3.

The actual values of vegetarians to meat-eaters = (3 × 3 parts) to (3 × 5 parts).

Ratio of vegetarians to meat-eaters = 9 : 15.

Of these 9 vegetarians, ⅔ eat fish.

$$\frac{2}{3} \times 9 = 6$$

So 6 out of the 9 vegetarians eat fish and therefore the remaining 3 do not eat fish.

Ratio of vegetarians to vegetarian fish-eaters to meat-eaters = 3 : 6 : 15.

Divide each part by 3 to express the ratio in its simplest terms = 1 : 2 : 5.

Q10 Answer = 17 : 28

You are asked to find a 'part to whole' ratio, where the 'part' is 'tea' and the 'whole' is 'all drinks'.

First multiply the ratio of 1 to 6 by a factor which will allow you to divide easily by 3 and 4. By multiplying each part of the 'part to part' ratio (1 : 6) by 4, the result is 4 cups : 24 mugs. So there are part (4 cups) + part (24 mugs) = whole (28 drinks).

You know that ¼ of the cups are filled with tea:

$$\frac{1}{4} \times 4\ cups = 1$$

You also know that ⅔ of the mugs are filled with tea:

$$\frac{2}{3} \times 24\ mugs = 16$$

So there are 17 tea drinks on the table out of a possible 28 drinks. So the ratio of tea to all drinks = 17 : 28.

Q11 Answer = 1375 miles

If ½ cm = 250 miles, 1 cm = 500 miles.

$$500 \times 2\frac{3}{4} = 1375$$

Q12 Answer = £180

Anne's 5 parts equal £225. 1 part is therefore £225 ÷ 5 = £45. Paula therefore receives 4 × £45 = £180.

Q13 Answer = 15 games

The ratio of games won to games lost:

$$\frac{Won}{Lost} = \frac{2}{3}$$

$$\frac{Actual\ games\ won}{Actual\ games\ lost} = \frac{(2 \times 3) = 6}{(3 \times 3) = 9}$$

Total games = 6 + 9 = 15 games.

Q14 Answer = 2 hours

If Charlotte ran the first half of the marathon in $\frac{4}{5}$ (time) of the second half, then the second half took her $\frac{5}{5}$ (time) of the second half. The total time is therefore:

$$\frac{4}{5} + \frac{5}{5} = \frac{9}{5}$$

9/5 = 4 hours and 30 minutes (270 minutes).

The ratio time of first half to total race = 4 : 9.

Charlotte's first half time is:

$$4 \times \left(\frac{270}{9}\right) = 120 \textit{ minutes}$$

Ignore the information about distance. It is not relevant to the question.

Q15 Answer = 240 : 1

You are given two 'part to part' ratios and asked to determine a relationship between two different parts. Set up the two part to part ratios:

Ratio 1	Ratio 2
Paper clips to Pencil sharpeners	Pencil sharpeners to Staplers
400 to 1	3 to 5

To compare the two ratios you need to find a common term for both ratios. By multiplying each part of ratio 1 by a multiple of 3 you create a common term in both ratios for the sharpeners.

Ratio 1			**Ratio 2**		
Paper clips to		Pencil sharpeners	Pencil sharpeners to Staplers		
400	to	1	3	to	5
1200	to	3	3	to	5
(400 × 3)	(1 × 3)				

Now you can compare Paper clips to Staplers: 1200 to 5. In its lowest terms, the ratio is 240 to 1.

Proportions practice drill 1

Q1 Yes. $3 \times 21 = 7 \times 9$
Q2 No. $2 \times 7 \neq 5 \times 3$
Q3 Yes. $2 \times 10 = 5 \times 4$
Q4 Yes. $12 \times 23 = 92 \times 3$
Q5 No. $5 \times 65 \neq 27 \times 12$
Q6 Yes. $36 \times 9 = 54 \times 6$
Q7 No. $3 \times 72 \neq 19 \times 12$
Q8 No. $2 \times 41 \neq 7 \times 12$
Q9 Yes. $7 \times 24 = 12 \times 14$
Q10 No. $4 \times 95 \neq 124 \times 3$

Proportions practice drill 2

Q1 $x = 2$
Q2 $x = 486$
Q3 $x = 12$
Q4 $x = 1$
Q5 $x = 10$
Q6 $x = 5$
Q7 $x = 49$
Q8 $x = 27$
Q9 $x = 78$
Q10 $x = 21{,}812$

Proportions practice questions

Q1 6 teachers
Q2 45 students
Q3 720 pixels
Q4 $123.75
Q5 $x = 6$
Q6 £2.25
Q7 $M = \frac{S}{6}$
Q8 $4x : 1$
Q9 20 carers
Q10 40 mountain bikes and 800 road bikes

Proportions practice questions explanations

Q1 Answer = 6 teachers

Set up a proportion and solve for the missing term:

$$\frac{4\ teachers}{24\ children} = \frac{t\ teachers}{36\ children}$$

24 children × t teachers = 4 teachers × 36 children
$24t = 144$

$$t = \frac{144}{24}$$

$t = 6$ teachers

Q2 Answer = 45 students study Chinese

You are given a 'part to part' ratio: there are 5 studying Chinese for every 4 studying Russia. Set up a proportion and solve for the number of students studying Chinese.

$$\frac{5\ Chinese}{9\ Chinese\ \&\ Russian} = \frac{C\ Chinese}{81\ Chinese\ \&\ Russian}$$

Cross-multiply and solve for C:
$9C = 5 \times 81 = 405$
$C = 45$
There are 45 students studying Chinese.

Q3 Answer = 720 pixels

Set up a proportion and solve for the width w:

$$\frac{180\ height}{240\ width} = \frac{540\ height}{w\ width}$$

$180w = 240 \times 540 = 129{,}600$
$w = 720$

Q4 Answer = $123.75

Set up a proportion and solve for x:

$$\frac{4\ keyboards}{\$99} = \frac{5\ keyboards}{x}$$

$4x = 5 \times \$99 = \$495$
$x = 123.75$
The unit is $, so the answer is $123.75.

Q5 Answer = 6

Find the missing term in the proportion $3 : 9 = x : 18$
Set up the ratios as fractions and solve for the missing term:

$$\frac{3}{9} = \frac{x}{18}$$

$3 \times 18 = 9x = 54$
$x = 6$

You can use common sense to solve this more quickly than setting up the equation. You can see that the second term in the second ratio is twice the second term in the first ratio ($18 = 2 \times 9$). So you could multiply the first term in the first ratio by two (3×2) and keep the ratios in proportion.

Q6 Answer = £2.25

Set up the ratios as a proportion and solve for the missing term:

$$\frac{12}{1.80} = \frac{15}{x}$$

$12x = 27$
$x = 2.25$
The unit is £, so the answer = £2.25.

Q7 Answer $M = \frac{S}{6}$

Set up a proportion with the known values:

$$\frac{10}{60} = \frac{\text{M minutes}}{\text{S sit ups}}$$

$60\text{ M} = 10\text{S}$

Divide both sides by 10.

$6\text{M} = \text{S}$

Divide both sides by 6.

$$M = \frac{S}{6}$$

Q8 Answer = $4x$: 1

Set up the ratio as a fraction and reduce the answer to its lowest terms:

$$\frac{16x^2}{4x^2}$$

Divide both sides by $4x$.

$$\frac{4x}{1} = 4x : 1$$

Q9 Answer = 20 carers

Let C be the current number of carers and 6C the current number of residents. This satisfies the ratio 1 : 6. In the future there will be 6C + 40 residents and 8 times as many residents as carers, so now you can solve for C:

$6C + 40 = 8C$
$40 = 8C - 6C$
$40 = 2C$
$C = 20$

There are 20 carers currently.

Q10 Answer = 40 mountain bikes and 800 road bikes

Let the number of mountain bikes = M and the number of road bikes = 20M. This satisfies the ratio 1 : 20.

M + 20M = Total number of bikes in the race.

If 200 road bikes were to join the race, the total number of road bikes would be 20M + 200.

When 200 road bikes have joined the race, there will be 25 times as many road bikes as mountain bikes:

25M = 20M + 200

Now you can solve for M:

20M + 200 = 25M
200 = 25M – 20M
200 = 5M
M = 40

40 is the constant number of mountain bikes in the race. Recall the ratio, mountain bikes to road bikes = 1 : 20. If there are 40 mountain bikes, you can find the number of road bikes by setting up a proportion:

$$\frac{1}{20} = \frac{40}{R}$$

Cross-multiply and solve for R:

1R = 20 × 40
R = 800
Mountain bikes = 40 and Road bikes = 800.

6

Data interpretation

Now you have had the opportunity to refresh your memory on the bulk of the skills you need to take a numerical reasoning test, in this chapter you have the opportunity to put them all together and use them to reason with data in charts, tables and graphs. This type of question tests not just your ability to perform rapid calculations under time pressure, but also your ability to think logically and identify exactly what the question is asking you. Test-writers will give you traps to fall into and lead you straight to a wrong answer in the choices, so beware of what is being asked of you and take a couple of extra seconds to re-read each question. Make absolutely sure you understand the question before attempting to solve the problem.

In this chapter, unlike in the preceding chapters, the answers are multiple choice. The reason is that typically you will be given a set of answer choices to choose from in numerical reasoning tests. A key skill in successful test-taking is the ability quickly to recognize any outliers in the answers. Eliminate these answers immediately, so that even if you end up guessing the wrong answer, you will at least reduce the probability of guessing from a wider range of incorrect answers.

In the following problems, you will use your knowledge of ratios, fractions, decimals, proportions and percentages. Try to complete each question within 5 minutes.

Tip: Where the answer choices are narrow in range, realize that you will have to work out the answer systematically. Where the answers are very wide apart, first eliminate the outliers, second estimate the correct answer, and third, pick the answer choice nearest to your estimate.

Data interpretation questions

1 Holiday insurance claims

Table 6.1 Data on Company Z insurance claims

Year	Total number of claims	Approx change on previous year	Total number of approved claims	Total number of non-approved claims
1993	966	15% increase	720	246
1994	1047	8.4% increase	813	?
1995	1013	3.2% decrease	726	?
1996	930	8.2% decrease	310	?
1997	975	4.8% increase	428	547

Q1 In which year was the greatest percentage change in the total number of claims on the previous year?
a) 1993 b) 1994 c) 1995 d) 1996 e) 1997

Q2 How many more claims were made in 1997 than in 1992?
a) 821 b) 135 c) 124 d) 174 e) 1,110

Q3 In which year did the number of non-approved claims exceed the number of approved claims in the ratio of 2 : 1?
a) 1993 b) 1994 c) 1995 d) 1996 e) 1997

Q4 The total number of claims made in 1997 is approximately what percentage change on the 1994 total number of claims?
a) 2% decrease b) 7% decrease c) 7% increase d) 15% decrease e) 40% increase

2 Goe-Ezy-Bizz flight charges

Table 6.2

(a) Airline flight charges

Outbound airport	Outbound tax	Outbound insurance	Inbound insurance	Inbound tax
Belfast	£5	£1.60	£2.60	£5
Edinburgh	£5	£1.60	£1.60	£3.23
Gatwick	£20	£0	£1.60	£3.23
Liverpool	£0	£1.60	£3.20	£0
Luton	£5	£3.20	£1.60	£0

(b) Fare schedule to Barcelona (excluding taxes and insurance)

Airport	Single	Return
Belfast	£15	£28.23
Edinburgh	£20	–
Gatwick	£25	£42.50
Liverpool	£20	£35
Luton	£20	£35

Q1 The cost of the return fare from Edinburgh to Barcelona is 140% of the single fare. What is the approximate total cost of the return airfare, including all taxes and charges, from Edinburgh to Barcelona?
a) £28 b) £48 c) £60 d) £40 e) £23

Q2 What is the difference between the price of a single flight to Barcelona from Liverpool and from Luton?
a) £0 b) £9.80 c) £6.60 d) £16.40 e) £19.20

Q3 Goe-Ezy-Bizz runs a late season summer sale, where prices are discounted by 7.5%. Which airport can offer the cheapest return flight to Barcelona, exclusive of taxes and insurance?
a) Belfast b) Edinburgh c) Gatwick d) Liverpool e) Luton

Q4 If the exchange rate of £ Sterling to euros is 1 : 1.55, how much does a single flight from Liverpool to Barcelona cost, exclusive of all additional charges?
a) 30 euros b) 31 euros c) 32 euros d) 35 euros e) 40 euros

3 Council services employment

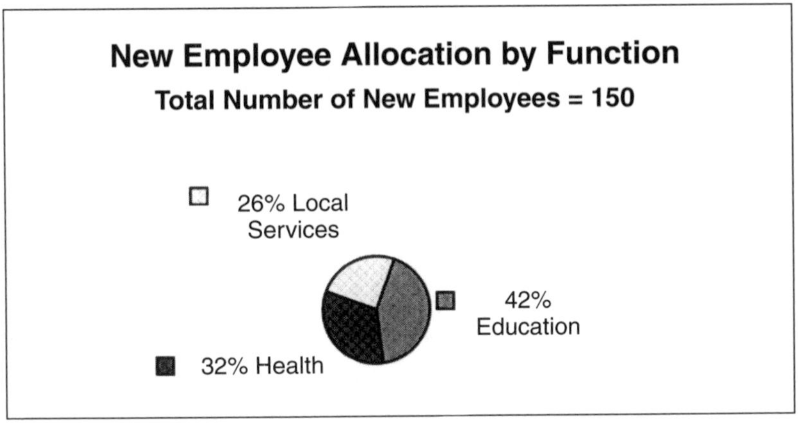

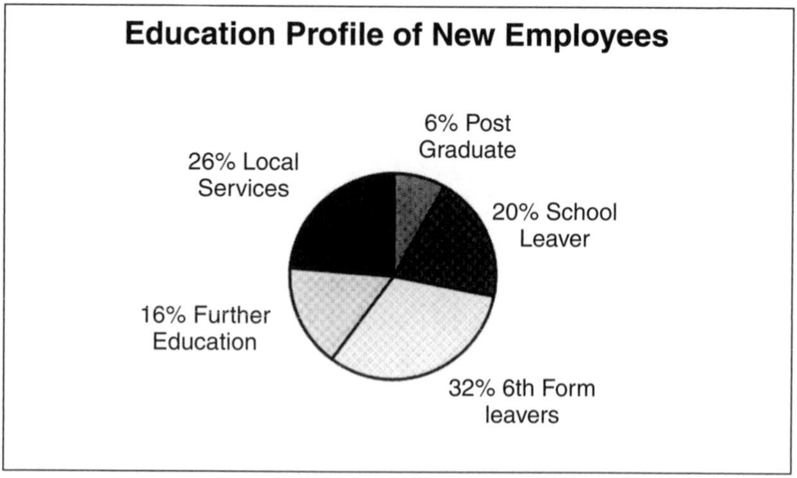

Figure 6.1

Q1 How many more people were hired into Education than into Health?
a) 15 b) 18 c) 21 d) 25 e) 26

Q2 If one-third of all new employees hired into Local Services were school leavers, how many school leavers were employed in Health and Education?
a) 13 b) 17 c) 30 d) 37 e) Cannot tell

Q3 How many sixth form leavers were employed in Local Services if the ratio of Education to Local Services to Health is 1 : 3 : 4 in the sixth form leavers category?
a) 6 b) 12 c) 18 d) 24 e) Cannot tell

Q4 Following a review, an additional six further education candidates were employed and allocated equally between the functions. This represents approximately what increase on the original number of further education candidates employed?
a) 12.5% b) 25% c) 33⅓% d) 40% e) 42%

4 European Union institutions 1995

Table 6.3 European Union institutions

Country	Population (millions)	Votes in Council of Ministers	Seats in European Parliament
Germany	80.6	10	99
France	57.5	10	87
Spain	39.1	8	64
Ireland	3.6	3	15
Luxembourg	0.4	2	6

Q1 A person from which country is best represented in the Council of Ministers?
a) Germany b) France c) Spain d) Ireland
e) Luxembourg

Q2 Which country has the least number of representatives in Parliament relative to its population size?
a) Germany b) France c) Spain d) Ireland
e) Luxembourg

Q3 If the ratio of population to votes were the same in France as in Ireland, how many more votes would France be entitled to in the Council of Ministers?
a) 23 b) 38 c) 48 d) 50 e) 110

5 Computer failure

Table 6.4 Lost revenue for Company X due to server downtime

Department	Average weekly server downtime	Total weekly lost revenue due to server downtime	Number of employees in department
A	4 hours	£225	24
B	6 hours	£288	18
C	2.5 hours	£155	30
D	1.5 hours	£78	1

Q1 On average, which department loses the most revenue per hour due to server downtime?
a) Dept A b) Dept B c) Dept C d) Dept D e) Cannot tell

Q2 Which department loses the most productive time per employee per week due to server downtime?
a) Dept A b) Dept B c) Dept C d) Dept D e) Cannot tell

Q3 Repair work is carried out which reduces the average weekly downtime in all departments by 25%. How many hours are gained following the repair work across all departments?
a) 3.5 b) 4.5 c) 8 d) 10.5 e) 12.75

6 Survey of voting turnout in 1994 local elections

Table 6.5 Percentage of voters by age category in the 1994 local elections

Constituency	18–24	25–34	35–44	45–54	55–74	75+	% of adults who did not vote	Total no. adults eligible to vote
Runneymede	2	14	19	19	18	2	26	14,650
Chudleigh	3	11	18	14	14	3	37	20,000
Bishopton	4	11	16	22	12	4	–	25,100
Thundersley	6	12	13	13	16	4	36	22,397

Q1 Approximately how many people eligible to vote in the 1994 election did not turn out in Bishopton?
a) 4,000 b) 7,500 c) 10,000 d) 12,500 e) 14,000

Q2 What percentage of total eligible adults in the 18–24 age category turned out to vote in Thundersley?
a) 3.8% b) 6% c) 9.5% d) 12.8% e) 15%

Q3 The 35–44 age group is reclassified in Chudleigh. As a result, 4% of the existing 35–44 age group are reclassified. How many people in Chudleigh does this affect?
a) 98 b) 144 c) 320 d) 450 e) 480

Q4 The number of voting adults in the 56–74 category as a proportion of all adults eligible to vote was greatest in which constituency?
a) Runneymede b) Chudleigh c) Bishopton
d) Thundersley e) Cannot tell

7 Mobile phone sales

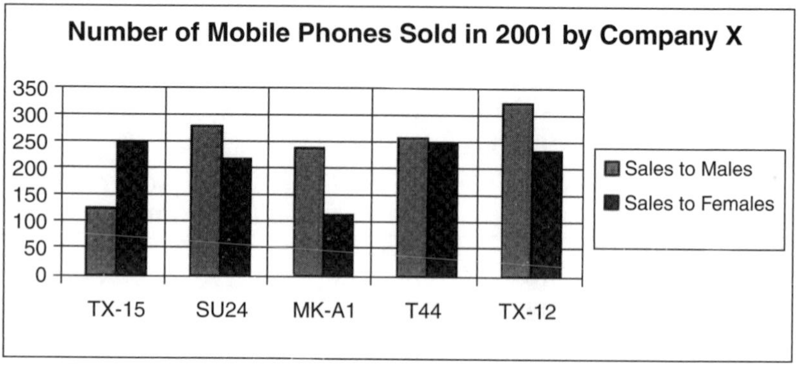

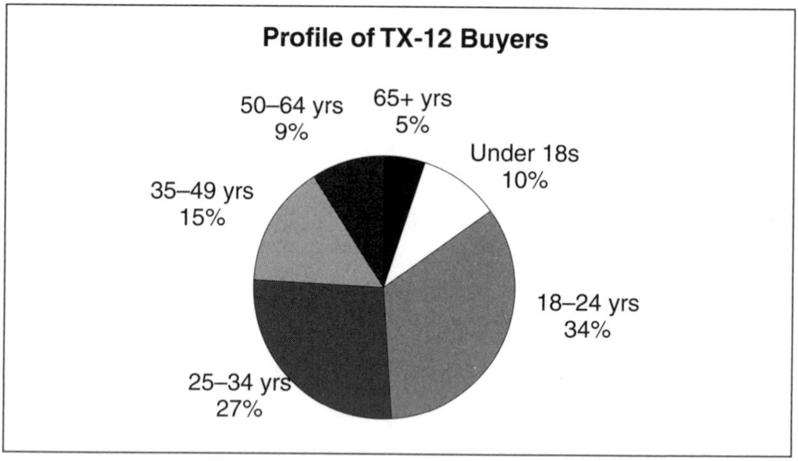

Figure 6.2

Q1 Approximately what percentage of all TX-15 sales were made to females?

a) 12.5% b) 31% c) 50% d) 66.6% e) 80%

Q2 Which make of mobile phone has the lowest ratio of male to female sales?

a) TX-15 b) SU24 c) MK-A1 d) T44 e) TX-12

Q3 Approximately how many of the TX-12 buyers are in the 35–49 age range?
a) 28 b) 32 c) 48 d) 85 e) 96

Q4 In 2002, sales projections indicate that sales of MK-A1 mobile phones will decline by 20%. Sales made to males and females are projected to decline proportionately. How many MK-A1 are predicted to be sold to females in 2002?
a) 82 b) 86 c) 96 d) 104 e) 112

8 Adult television viewing hours

Table 6.6 Adult viewing preferences (%)

	BBC1		BBC2		ITV		Channel 4		Total viewing hrs per week	
	M	F	M	F	M	F	M	F	M	F
1984	12	11	8	6	20	21	12	10	12.2	8.5
1985	13	10	9	10	18	19	9	12	14.3	8.2
1986	13	12	12	9	20	14	12	8	14.7	8.4
1987	14	12	10	9	18	15	13	9	15.2	9.0
1988	18	12	6	9	18	18	11	8	15.6	9.1

Q1 What was the approximate average weekly viewing time for females watching the BBC in 1984?
a) 2 hours 20 minutes b) 88 minutes c) 1 hour 15 minutes d) 45 minutes e) Cannot tell

Q2 In which year was the ratio of male to female average viewing time the lowest?
a) 1984 b) 1985 c) 1986 d) 1987 e) 1988

Q3 What was the approximate total average female viewing time in 1983, if the average female viewing time declined by 15% between 1983 and 1984?

a) 9.5 hours b) 9.7 hours c) 10.0 hours
d) 10.5 hours e) 11.0 hours

Q4 Approximately how many more men than women on average watched television on Mondays in 1987?

a) 16 b) 6 c) 84 d) 96 e) Cannot tell

9 McCoopers Consultancy New York Office

Application categories	Number of applicants		Number of job offers made	
	1999	2000	1999	2000
MBA postgraduates	150	200	20	60
Industry specialists	78	112	30	60
Other consultancies	42	76	7	4
Academia	24	18	9	6

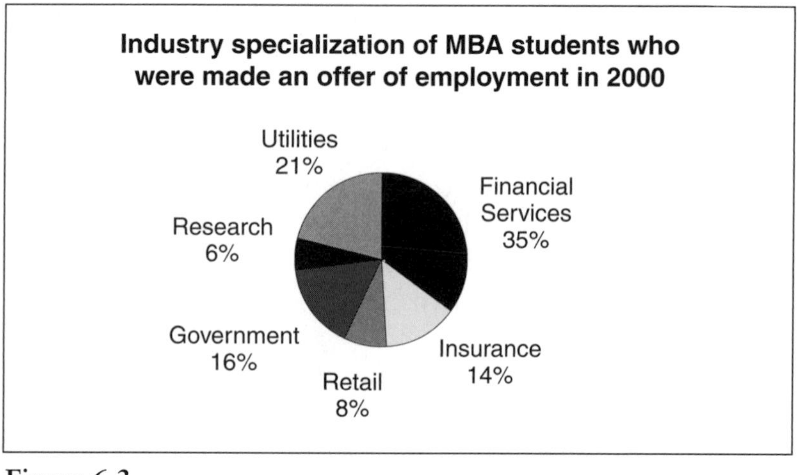

Figure 6.3

Q1 Which group had the lowest ratio of job offers to applicants in 1999?
a) MBA b) Industry specialists c) Other consultancies
d) Academia

Q2 Which group showed the largest percentage change in applications between 1999 and 2000?
a) MBA b) Industry specialists c) Other consultancies
d) Academia

Q3 Of those MBA postgraduates who were made an offer of employment in 2000, how many were financial services specialists?
a) 16 b) 21 c) 30 d) 32

Q4 In which of the following groups was the greatest number of job offers made per applicant?
a) MBA 2000 b) Other consultancies 1999
c) Other consultancies 2000 d) Academia 2000

10 Consultancy rates

Table 6.7 Hirst Consulting charge-out costs

Consultant	Daily rate (£)	Mileage home to client	Charge travel time?	Charge overtime?
Clancy	240	40 miles	Yes	Yes
Mellor	260	35 miles	Yes	Yes
Osborne	350	70 miles	No	No
Smith	400	20 miles	No	Yes

Consultants from Hirst Consulting charge their clients a fixed daily rate for an 8-hour day and expenses according to the consultant's agreed contract. Where applicable, overtime is charged at 50% of the pro rata hourly rate and 1 hour pro rata standard rate is charged for daily travel, regardless of the distance or time travelled. Consultants may claim 8 p per mile from Hirst Consulting for petrol expenses and do not charge the client for this cost.

Q1 What is the approximate cost to the client to have Clancy and Osborne on site for a 5-day project, where Clancy works an average 9.5 hours per day and Osborne works the minimum 8 hours per day?
a) 2,800 b) 3,200 c) 3,600 d) 4,000 e) Cannot tell

Q2 Mellor is on the client site for 22 days in January and February. How much does he charge to Hirst Consulting for petrol?
a) £154 b) £142.50 c) £123.20 d) £112.60 e) Cannot tell

Q3 Following a promotion, Smith's daily rate increases by 15%. How much extra will the client have to pay per day to have Smith on the project following the increase?
a) £460 b) £400 c) £120 d) £60 e) Cannot tell

11 Revenues for Cookie's Bakery

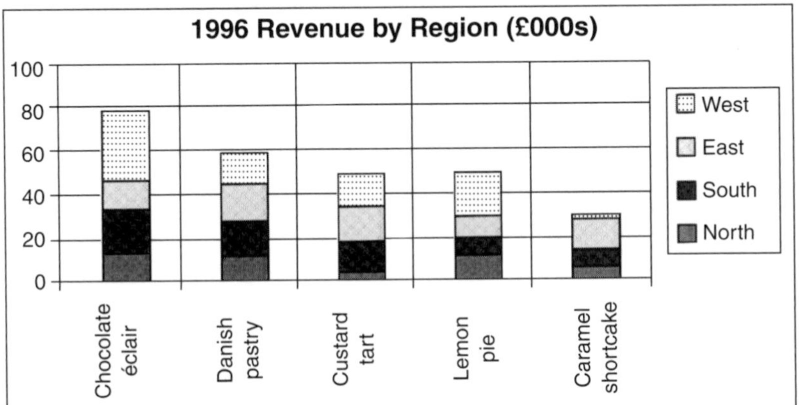

1996 Average profit margin (%) per cake

Caramel shortcake	19	Custard tart	9.5
Chocolate éclair	8	Danish pastry	12.5
Lemon pie	14		

Figure 6.4

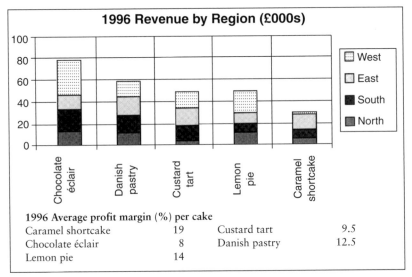

Q1 Revenues of Danish pastries in the West represent approximately what percentage of total Danish pastry revenue in 1996?

a) 12.5% b) 15% c) 22% d) 35% e) 40%

Q2 Approximately how much profit was made on custard tarts in the North?

a) £400 b) £800 c) £1,400 d) £1,250 e) £2,400

Q3 In 1997, revenues from all cakes in the range increased to £300,000. This represents an increase of what percentage on 1996 revenues?

a) 9% b) 10% c) 11% d) 12.5% e) 15%

Q4 In 1997, lemon pie profits increased to 20%. 1997 revenues from lemon pies increased by 2.5% on the previous year. How much profit was made on the lemon pie line in 1997?

a) £4,200 b) £5,125 c) £5,400 d) £10,250 e) £51,250

12 Currency fluctuation

Table 6.8 Great Britain Sterling exchange rates

	May	June	July	August
Euro	1.44	1.45	1.48	1.50
US Dollar	1.60	1.60	1.61	1.53
Russian Rouble	47.31	48.22	48.42	62.31
Japanese Yen	179.20	179.21	177.66	182.00
Slovakian Koruna	57.34	57.68	61.34	69.07

Q1 Which currencies increased in price against the GB £ in any month in the May to August period?

a) Dollar and rouble b) Dollar and yen c) Yen and koruna
d) Euro and dollar e) Euro and koruna

Q2 If the euro weakened against the GB £ by 6% between August and September, approximately how many euros could I buy for £25 in September?

a) 1.25 euros b) 60 euros c) 40 euros
d) 85 euros e) 35 euros

Q3 Which currency showed the largest fluctuation against the GB £ in the period May to August?

a) Euro b) Dollar c) Rouble d) Yen e) Koruna

Q4 If the Slovakian koruna increased in the same proportion as the Japanese yen between August and September, and the yen exchange rate was 173 yen to the GB £ in September, approximately how many koruna were there to the £ in September?
a) 66 b) 73 c) 93 d) 112 e) 146

13 Gym membership

Table 6.9 Cost of gym membership Chewton Magna District

Gym	All-inclusive monthly membership	Minimum membership period	Price per class	End-of-year holiday closure
Doddington	£41	6 months	£4.50	14 days
Kearns	£37.50	6 months	£4.75	4 days
Hagen	£46	None	£4.20	1 day
Deane	£26	12 months	£2.75	3 days

Q1 If I want to pay for gym membership from 1 March to 31 July, which gym offers the best deal?
a) Doddington b) Kearns c) Hagen
d) Deane e) Cannot tell

Q2 If I attend seven classes per month for a year, which gym offers cheaper all-inclusive membership than a pay-per-class payment plan? (Ignore minimum membership.)
a) Doddington b) Kearns c) Hagen
d) Deane e) None of the above

Q3 Which gym is the most expensive in terms of a pro rata daily rate over an annual period?
a) Doddington b) Kearns c) Hagen
d) Deane e) Cannot tell

14 Kishbek Semiconductor sales

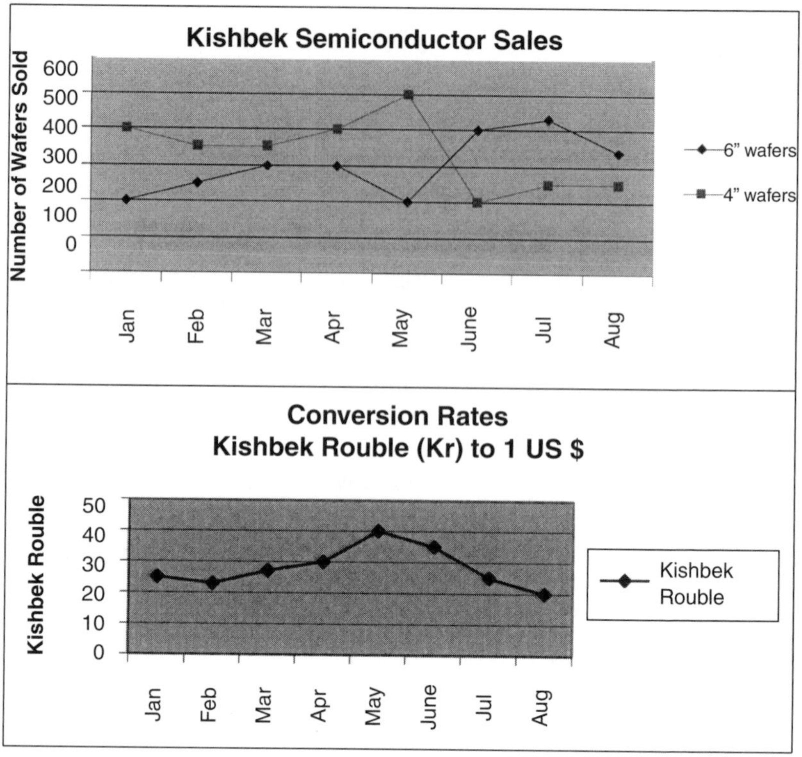

Figure 6.5

Q1 What was the buying cost of 100 4″ wafers in May if the total revenue received for 4″ wafers was 10,000 Kishbek roubles?

a) 2 roubles b) 20 roubles c) 200 roubles
d) 500 roubles e) 2000 roubles

Q2 What was the approximate percentage increase in the sales of 6″ wafers between May and June?

a) 50% b) 100% c) 150% d) 200% e) 400%

Q3 How much more would it cost a US $ purchaser to buy 250 6″ wafers in August than in May, if the selling price of 6″ wafers remains stable at 125 Kishbek roubles per five 6″ wafers?
a) $148.50 b) $156.25 c) $72.50 d) $68.40 e) $168.80

15 Energy tariffs

Table 6.10 Energy tariffs from North–South Power

Tariff	Energy supply	Service charge	Unit price
Tariff A	Gas only	9.99p per day	1.32p per kwh*
Tariff B	Electricity only	13.39p per day	
Tariff C	Gas & electricity	£11.50 per year*	**Gas** 1.25p per kwh **Electricity** 5.02p per kwh

*kwh = kilowatt hour
** Tariff C requires an upfront non-refundable full payment of the service charge.

Q1 What is the approximate cost to a customer for gas on tariff A for 2,854 kilowatt hours consumed over a 20-day period?
a) £20 b) £25 c) £30 d) £35 e) £40

Q2 Electricity purchased on tariff C is 2.5% cheaper than on tariff B. What is the approximate cost of 640 kwh of electricity purchased on tariff B and consumed over 91 days?
a) £45 b) £11 c) £67 d) £82 e) 101

Q3 What is the approximate difference in price between gas and electricity purchased separately on tariff A and tariff B and gas and electricity purchased on tariff C if 250 kwh of gas and 120 kwh of electricity are consumed over 15 days? (Use any relevant information from previous questions to answer Q3.)
a) £12.50 b) £10 c) £7.50 d) £3.33 e) £1.25

Answers to Chapter 6

1 Holiday insurance claims

Q1 a) 1993
Q2 b) 135
Q3 d) 1996
Q4 b) 7% decrease

2 Goe-Ezy-Bizz flight charges

Q1 d) £40
Q2 c) £6.60
Q3 b) Edinburgh
Q4 b) 31 euros

3 Council services employment

Q1 a) 15
Q2 b) 17
Q3 c) 18
Q4 b) 25%

4 European Union institutions 1995

Q1 e) Luxembourg
Q2 a) Germany
Q3 b) 38

5 Computer failure

Q1 c) Department C
Q2 b) Department B
Q3 a) 3.5 hours

6 Survey of voting turnout in 1994 local elections

Q1 b) 7,500
Q2 a) 3.8%
Q3 b) 144
Q4 a) Thundersley

7 Mobile phone sales

Q1 d) 66.6%
Q2 a) TX-15
Q3 d) 85
Q4 c) 96

8 Adult television viewing hours

Q1 e) Cannot tell
Q2 a) 1984
Q3 c) 10 hours
Q4 e) Cannot tell

9 McCoopers Consultancy New York Office

Q1 a) MBA
Q2 c) Other consultancies
Q3 b) 21
Q4 d) Academia 2000

10 Consultancy rates

Q1 b) £3,200
Q2 c) £123.20
Q3 d) £60

11 Revenues for Cookie's Bakery

Q1 c) 22%
Q2 a) £400
Q3 c) 11%
Q4 d) £10,250

12 Currency fluctuation

Q1 b) Dollar and yen
Q2 c) 40 euros
Q3 c) Rouble
Q4 a) 66

13 Gym membership

Q1 b) Kearns
Q2 e) None of the above
Q3 c) Hagen

14 Kishbek Semiconductor sales

Q1 e) 2000 roubles
Q2 b) 100%
Q3 b) $156.25

15 Energy tariffs

Q1 e) £40
Q2 a) £45
Q3 c) £7.50

Explanations to Chapter 6 questions

Section 1 Holiday insurance claims

Q1 a) 1993

You can read this information directly from the table. 1993 had the greatest percentage change on the previous year.

Q2 b) 135

Between 1992 and 1993, the number of claims increased by 15%, therefore (1993 value) plus 15% = 966.

$$966 \times \frac{100}{115} = 840$$

In 1992 there were 840 claims and in 1997 there were 975 claims. Therefore in 1997 there were (975 – 840) more claims than in 1992.

Q3 d) 1996

In order to find the number of non-approved claims, subtract the number of approved claims from the total number of claims.

Year	Total	Approved	Non-approved
1993	966	720	246
1994	1047	813	234
1995	1013	726	287
1996	930	310	620
1997	975	428	547

Now look for a ratio of non-approved claims to approved claims in the ratio of 2 : 1. There are only two years in which the number of non-approved claims exceeds approved claims, 1996 and 1997:

1996 ratio = 620 : 310
1997 ratio = 547 : 428

By reducing the ratios to the simplest form you will see that the 1996 ratio = 2 : 1.

Q4 b) 7% decrease
1997 total number of claims = 975; 1994 total number of claims = 1047. The number of claims in 1997 represents a decrease on the 1994 total, so eliminate all the answers that represent an increase. You are left with answer choices a), b) and d). Now work out the actual percentage decrease. Recall the formula for percentage change.

$$\%\ change = \frac{actual\ change}{original} \times 100\%$$

$$\%\ change = \frac{(1047 - 975)}{1047} \times 100\% \approx$$

$$\frac{70}{1050} \times \frac{100}{1} \approx \frac{7000}{1000} \approx 7\%$$

Section 2 Goe-Ezy-Bizz flight charges

Q1 d) £40
Single fare Edinburgh to Barcelona = £20

Return fare = 20 × 140% =	£28
Outbound insurance	£1.60
Inbound insurance	£1.60
Outbound tax	£5
Inbound tax	£3.23

Total = £39.43. The question asks for the approximate cost so choose the closest answer.

Q2 c) £6.60

	Liverpool	**Luton**
Fare	£20	£20
O/B insurance	£1.60	£3.20
O/B tax	£0	£5
Total	£21.60	£28.20

The difference in price = £28.20 – £21.60 = £6.60

Q3 b) Edinburgh
In Q1 you worked out the price of a return flight from Edinburgh (£28). As all the flights are discounted by the same percentage you can read directly from the table the price of the cheapest return fare without completing the calculation.

Q4 b) 31 euros
The price of a single fare from Liverpool to Barcelona = £20. To convert £20 to euros, multiply by the exchange rate. £20 × 1.55 euros = 31 euros.

Section 3 Council services employment

Q1 a) 15
In Education 42% were hired. In Health 32% were hired. The difference = 10%. 10% of 150 = 15.

Q2 b) 17
The total number hired into Local Services is:

$$26\% \times 150 = 150 \times \frac{26}{100} = 39$$

The question tells you that one-third of the total (39) hired into Local Services were school-leavers. Therefore ($\frac{1}{3} \times 39 = 13$) school leavers were hired into Local Services. Figure 6.1 tells you that 20% of the new employees were school leavers, and 20% × 150 = 30. If 13 were hired into Local Services, then (30 – 13) = the number hired into Education and Health = 17.

Q3 c) 18

The actual number of sixth form leavers employed in Local Services = 32% × 150.

$$150 \times \frac{32}{100} = 48$$

You are given a part : part : part ratio and the sum of the parts in a part : part ratio = the whole.

1 : 3 : 4 = 1 + 3 + 4 = 8. The proportion employed in local services is 3 parts of the whole, or $\frac{3}{8}$.

$\frac{3}{8}$ × the total number of sixth form leavers =

$$\frac{3}{8} \times 48 = 18$$

Q4 b) 25%

The actual number of further education candidates employed = 16% × total number employed:

$$\frac{16}{100} \times 150 = 24$$

You may recognize immediately that 6 is $\frac{1}{4}$ of 24, or 25%. If not, use the percentage change formula and plug in the numbers.

$$\%\ \mathit{change} = \frac{\mathit{actual\ change}}{\mathit{original}} \times 100\%$$

$$\% \ change = \frac{6}{24} \times 100\% = 25\%$$

Section 4 European institutions

Q1 e) Luxembourg

The question is asking 'how many people does 1 vote represent in the Council of Ministers?' You are really working out a 'part to part' ratio.

$$Germany = \frac{80.6\ m}{10\ votes} \approx \frac{80\ m}{10} \approx 8\ m/vote$$

$$France = \frac{57.5\ m}{10\ votes} \approx \frac{60\ m}{10} \approx 6\ m/vote$$

$$Spain = \frac{39.1\ m}{8\ votes} \approx \frac{40\ m}{8} \approx 5\ m/vote$$

$$Ireland = \frac{3.6\ m}{3\ votes} \approx 1.2\ m/vote$$

$$Luxembourg = \frac{0.4\ m}{2\ votes} \approx 0.2\ m/vote$$

In Luxembourg, 1 vote represents approximately 0.2 m people whereas in Germany 1 vote represents approximately 8 m people, so Luxembourg is best represented.

Q2 a) Germany

The question asks you which country is least represented by seats in the Parliament relative to its population size. Set up a 'part to part' ratio. This will tell you how many of the population are represented by each seat. Round the numbers up to ease the calculation.

	Seats	Population	Approximate ratio
Germany	99 (≈ 100)	80.6 m (≈ 80 m)	≈ 1 : 800,000
France	87 (≈ 90)	57.5 m (≈ 60 m)	≈ 1 : 700,000
Spain	64 (≈ 60)	39.1 m (≈ 40 m)	≈ 1 : 600,000
Ireland	15 (≈ 15)	3.6 m (≈ 4 m)	≈ 1 : 200,000
Luxembourg	6	0.4 m	≈ 1 : 100,000

Germany has the least representation in the Parliament.

Q3 b) 38

Population	Total votes	=	Citizens	:	1 vote
Ireland	3.6m	=	1.2m	:	1

In Ireland, 1.2 m are represented with one vote. If France had similar representation, it would also be entitled to 1 vote per 1.2 m. Total population in France = 57.5 m, so divide the total by 1.2.

$$\frac{57.5m}{1.2} \approx 48\ votes$$

France has 10 votes, so would be entitled to a further 38 votes.

Section 5 Computer failure

Q1 c) Department C

You are asked to work out how much revenue is lost per hour per department. Divide the total weekly lost revenue by the number of hours the server is down for each department.

$$A = \frac{225}{4} \approx 56 \quad B = \frac{288}{6} = 48 \quad C = \frac{155}{2.5} = 62$$

$$D = \frac{78}{1.5} = 52$$

Department C loses more per hour than the other departments.

Q2 b) Department B
In each department, each employee loses the amount of time lost by the department. Employees in Department B lose the most time (6 hours.)

Q3 a) 3.5 hours
The total amount of time lost on average per week = (4 + 6 + 2.5 + 1.5) = 14 hours

Now decrease the total lost hours by 25% to find the new total.
14 × 25% = 3.5 hours.

Section 6 Survey of voting turnout in 1994 local elections

Q1 b) 7,500
The percentage of those who didn't turn out to vote = (100% – total percentage of voting adults). In Bishopton (4 + 11 + 16 + 22 + 12 + 4) the percentage of adults who voted = 69%. So the percentage of adults who didn't vote = 31%. The question asks you for an approximate value, so find 30% of the total:

10% × 25,100 = 2,510 so 30% × 25,100 = (10% × 3) = 7,530.

Q2 a) 3.8%
The percentage of voting adults in the 18–24 category in Thundersley = 6%.
The percentage of all voting adults in Thundersley = 64%.
6% of 64% = 0.06 × 64 = 3.8% of eligible adults who voted were aged 18–24.

Q3 b) 144

You want to find 4% of 18% of the number of adults eligible to vote in Chudleigh.

To estimate the answer find 5% × 20% of 20,000.

$$\frac{5}{100} \times \frac{20}{100} \times 20{,}000 = 200$$

The answer choices closest to your estimate are b) 144 and c) 320. You can eliminate answers a), d) and e) at this point. To find the exact answer find 4% of 18% and multiply by 20,000:

$$\frac{4}{100} \times \frac{18}{100} \times 20{,}000 = 144$$

Q4 d) Thundersley

Write the percentage of voting adults in the 55–74 age category as a fraction of all voting adults in each constituency. With a rough estimate, you can see that Thundersley > Bishopton and Thundersley > Chudleigh, so eliminate Bishopton and Chudleigh. (If this doesn't seem obvious at first, use the cross-multiplying technique you learnt in Chapter 5.)

You are left with Thundersley and Runneymede. R = $\frac{18}{74}$ and T = $\frac{16}{64}$. Reduce both fractions to their lowest terms: R = $\frac{9}{37}$ and T = $\frac{1}{4}$. Use your knowledge of proportions to work out which is the larger of the two fractions:

	R	?	T	
(9 × 4 = 36)	$\frac{9}{37}$	?	$\frac{1}{4}$	(37 × 1 = 37)

Now ask 'is 36 > 37?' The answer is 'no', so Thundersley has the greatest proportion of voting adults in the 55–74 category.

Section 7 Mobile phone sales

Q1 d) 66.6%

Sales of TX-15 to males = 125
Sales of TX-15 to females = 250
Total sales of TX-15 = 375
Therefore, % sales to females =

$$\frac{(part)\ 250}{(whole)\ 375} = \frac{2}{3} \text{ or } 66.6\%$$

Q2 a) TX-15

Read off the graphs the ratios that tell you that fewer males bought phones than females and eliminate the rest. Only the TX-15 was sold to fewer males than to females.

Q3 d) 85

In Figure 6.2, read off the total number of TX-12 sales ≈ 325 + 240 = 565. Figure 6.2 tells you that 15% of the total TX-12 sales were made to 35–49 year olds.

$$\frac{15}{100} \times 565 \approx 85$$

Q4 c) 96

Total MK-A1 sales in 2001 = (240 + 120) = 360. The part to part ratio of male to female sales = 240 : 120 or 2 : 1. If sales decline by 20% in 2002:

$$360 \times \frac{80}{100} = 288 \text{ (total) will be sold}$$

The ratio of males to females remains constant at 2 : 1. Each part = one and there are three parts. To find the ratio 2 male : 1 female, divide the total by three to find the 1 part female.

$$\frac{288}{3} = \textit{sales to females}$$

Projected sales to females = 96.

Section 8 Adult television viewing hours

Q1 e) Cannot tell

The table does not tell you the actual number of television-watching adults, so you cannot work out the percentage of the total females.

Q2 a) 1984

Set up the male to female ratios for each year as fractions and round the numbers to help you to reduce the fractions to the simplest form:

$$1984 \approx \frac{12}{9} \left(= \frac{4}{3}\right) \quad 1985 \approx \frac{14}{8} \left(= \frac{7}{4}\right)$$

$$1986 \approx \frac{15}{8} \left(= \frac{15}{8}\right) \quad 1987 \approx \frac{15}{9} \left(= \frac{5}{3}\right)$$

$$1988 \approx \frac{16}{9} \left(= \frac{16}{9}\right)$$

Some of the fractions have the same denominator, so you can easily make a comparison:

$$1984\ \frac{4}{3} < 1987\ \frac{5}{3}$$

Eliminate 1987 as the higher ratio:

$$1987\ \frac{15}{9} < 1988\ \frac{16}{9}$$

Eliminate 1988 as the higher ratio:

$$1987 = \frac{15}{9} < 1986 \ \frac{15}{8}$$

Eliminate 1986 as the higher ratio. You are left with two answer choices to compare. Compare 1984 and 1985 with a proportion:

$$(1984: 4 \times 4 = 16) \quad \frac{4}{3} \qquad ? \qquad \frac{7}{4} = (1985: 3 \times 7 = 21)$$

1984 is the lower ratio of male to female viewing time.

Q3 c) 10 hours
The average female viewing time in 1984 = 8.5 and the 1983 value is 15% lower than the 1984 value. Thus the 1984 figure represents 85% of the 1983 figure (x):

$$8.5 = \frac{85}{100} \times x$$

Rearrange the formula to find x:

$$x = \frac{85 \times 100}{85} = \frac{850}{85} = 10 \textit{ hours}$$

Q4 e) Cannot tell
You are not given the actual total number of adults watching television, so you cannot work out the percentage of an unknown total. Remember that to work out the actual number that a percentage represents, you need to know the total number representing 100%.

Section 9 McCoopers Consultancy New York Office

Q1 a) MBA

Set up the ratio of offers made to number of applicants then reduce the fractions to their simplest terms:

$$MBA = \frac{20}{150} = \frac{2}{15} \quad Industry\ specialists = \frac{30}{78} = \frac{5}{13}$$

$$Other\ consultancies = \frac{7}{42} = \frac{1}{6} \quad Academia = \frac{9}{24} = \frac{3}{8}$$

Now arrange the ratios in size order.

$$\frac{2}{15} < \frac{1}{6} < \frac{3}{8} < \frac{5}{13}$$

The smallest ratio is the MBA group.

Q2 c) Other consultancies

Recall the formula for a percentage change. Make an estimate of the two largest percentage changes and plug in the numbers:

$$\%\ change = \frac{actual\ change}{original} \times 100\%$$

$$Industry\ specialists\ \%\ change = \frac{(112 - 78)}{78} \times 100\% \approx$$

$$\frac{17}{39} \times \frac{100}{1} \approx \frac{1}{2} \times \frac{100}{1} \approx 50\%$$

Industry specialists % change = 50%.

$$Other\ consultancies\ \%\ change = \frac{(76 - 42)}{42} \times 100\% \approx$$

$$\frac{34}{42} \times \frac{100}{1} \approx \frac{3}{4} \times \frac{100}{1} \approx 75\%$$

Other consultancies percentage change = 75%. Therefore, the 'Other consultancies' category showed the largest percentage change.

Q3 b) 21
Refer to the top half of Figure 6.3 to find the number of MBA offers in 2000 = 60 and the bottom half of the figure to obtain the percentage of MBA graduates with a preference to work in the financial services sector = 35%.

$$60 \times \frac{35}{100} = 21$$

Q4 d) Academia 2000
Set up a ratio for the number of offers made to the number of applicants and reduce the ratio to its lowest terms:

$$MBA\ 2000 = \frac{60}{200} = \frac{3}{10}$$

$$\textit{Other consultancies}\ 1999 = \frac{7}{42} = \frac{1}{6}$$

$$\textit{Other consultancies}\ 2000 = \frac{4}{76} = \frac{1}{19}$$

$$\textit{Academia}\ 2000 = \frac{6}{18} = \frac{1}{3}$$

Now arrange the fractions in size order.

$$\frac{1}{3} > \frac{3}{10} > \frac{1}{6} > \frac{1}{19}$$

Academia 2000 has the largest ratio of job offers per applicant. If you are in doubt as to the relative size of the fractions, use a proportion to work it out. For example, which is larger, $\frac{3}{10}$ or $\frac{6}{18}$?

$$(18 \times 3 = 54) \qquad \frac{3}{10} \quad ? \quad \frac{6}{18} \qquad = (10 \times 6 = 60)$$

Now ask 'is 54 > 60?' The answer is no, so you know that $\frac{6}{18}$ is the larger ratio.

Section 10 Consultancy rates

Q1 b) £3,200

Add all the relevant costs to the client per consultant.

	Daily rate	Overtime	Travel
Clancy	£240 × 5 days	7.5 hours (50% × £30)	5 hours × £30 = £1,462.50
Osborne	£350 × 5 days	n/a	n/a = £1,750

Total charged to client = £1,462.50 + £1,750 = £3,212.50.

Q2 c) £123.20

22 days × 70 miles round trip × 8p per mile = £123.20.

Q3 d) £60

Find 15% of Smith's daily rate:

$$400 \times \frac{15}{100} = £60$$

The client will have to pay an extra £60 per day.

Section 11 Revenues for Cookie's Bakery

Q1 c) 22%

Recall the formula to find a percentage:

$$\frac{Part}{Whole} = Percentage$$

$$\frac{13}{60} \approx 25\%$$

Choose the answer closest to 25%.

Q2 a) £400

Custard tart profit = 9.5%.

Revenues for custard tarts in the North = £4,000.

Profit = 9.5% × £4,000 ≈ 10% of £4,000 ≈ £400.

Q3 c) 11%

Total revenues on all cakes in 1996 = (80 + 60 + 50 + 50 + 30) = 270,000.

Total revenues on all cakes in 1997 = 300,000.

Recall the formula to find the percentage change:

$$\%\ change = \frac{actual\ change}{original} = \frac{30{,}000}{270{,}000} = \frac{1}{9} = approx\ 11\%$$

Q4 d) £10,250

The question tells you that in 1997 lemon pie profits = 20% and that 1997 revenues showed a 2.5% increase on 1996 lemon pie revenues.

1997 lemon pie revenue = £50,000 × 1.025 = £51,250.

Therefore, 1997 lemon pie profit = £51,250 × 20%.
10% of £51,250 = £5,125 (divide £51,250 by 10) and therefore 20% = £5,125 × 2 = £10,250.

Section 12 Currency fluctuation

Q1 b) Dollar and yen
Look for currencies that increase in value relative to the GB £. (This means that you get less currency for one GB £). The US $ strengthens against the GB £ between July and August (1.61 increases to 1.53). The Japanese yen also strengthens against the GB £ between June and July (179.21 increases to 177.66).

Q2 c) 40 euros
If the euro fell by 6%, you will receive 6% more currency for every GB £. First calculate the value of the euro in September by multiplying the value of the euro by 6%:

1.5 × 1.06 = 1.59

So in September 1.59 euros can be exchanged for £1, so £25 will buy 1.59 euros × £25 = 39.75 euros, or approximately 40 euros.

Q3 c) Rouble
Look for the currency that showed the largest percentage change in the period. First eliminate the answer choices that show an obviously small percentage change (euro, dollar and yen). Now work out the percentage change of the remaining choices. Use the percentage change formula:

$$\%\ \textit{change} = \frac{\textit{actual change}}{\textit{original whole}} \times 100\%$$

Rouble

$$\% \ change = \frac{(62.31 - 47.31)}{47.31} \times 100\% \approx$$

Round the numbers to estimate the answer:

$$\frac{62 - 47}{47} \times \frac{100}{1} \approx$$

Complete the calculation:

$$\frac{15}{47} \times \frac{100}{1} \approx \frac{1}{3} \times \frac{100}{1} \ or \ 33.3\%$$

Koruna

$$\% \ change = \frac{(69.07 - 57.34)}{57.34} \times 100\%$$

Round the numbers to estimate the answer:

$$\frac{69 - 57}{57} \times \frac{100}{1} \approx$$

Complete the calculation:

$$\frac{12}{60} \times \frac{100}{1} \approx \frac{1}{5} \ or \ 20\%$$

Therefore, the rouble shows the largest percentage change.

Q4 a) 66

Using the percentage change formula work out the percentage change for the yen between August and September:

$$\frac{182 - 173}{182} \times 100\% =$$

$$\frac{9}{182} \times \frac{100}{1} \approx \frac{1}{20} \times \frac{100}{1} \approx \frac{100}{20} \textit{ or } 5\%$$

Now increase the koruna by the same amount. Work out 5% of 69:

$$\frac{69}{1} \times \frac{5}{100} \approx 70 \times \frac{5}{100} \approx \frac{350}{100} = 3.5$$

Subtract 5% from the value of the koruna in August: 69 – 3.5 = 65.5. The question asks for an approximate answer, so select the answer closest to 65.5. Remember in currency conversion questions that when a currency *increases or strengthens* against the GB £, the amount of currency you receive for each pound *decreases*. The opposite is true when a currency weakens or falls.

Section 13 Gym membership

Q1 b) Kearns

Quickly calculate the cost of 5 months membership at each gym, remembering the minimum requirements for membership.

Doddington	41 × 6	= 246
Kearns	37.50 × 6	= 225
Hagen	46 × 5	= 230
Deane	26 × 12	= 312

Even though Kearns requires a minimum 6-month membership, it is still less expensive to pay for 6 months at Kearns than 5 months at Hagen, which is the only gym to offer membership for less than 6 months.

Q2 e) None of the above

Gym	Pay-per-class × 7	Membership
Doddington	£4.50 × 7 = £31.50	£41
Kearns	£4.75 × 7 = £33.25	£37.50
Hagen	£4.20 × 7 = £29.40	£46
Deane	£2.75 × 7 = £19.25	£26

All the gyms are more expensive for monthly membership.

Q3 c) Hagen

Kearns and Deane are obviously less expensive than Doddington and Hagen, so eliminate these first. Hagen is more expensive per month than Doddington, but Doddington is closed for more days in the year, which increases the daily pro rata rate.

The question asks you for an approximate answer, so make a rough estimate of the correct answer by working out the total annual price of gym membership and divide by the number of days the gym is open in the year.

$$Doddington = \frac{\text{£}492}{351 \text{ days}} < 1.5$$

$$Hagen = \frac{\text{£}552}{364 \text{ days}} \approx 1.5$$

Doddington costs less than Hagen, so Hagen is the most expensive.

Section 14 Kishbek Semiconductor sales

Q1 e) 2000 roubles

In May, 500 4″ wafers were sold. Total revenue for 4″ wafers was 10,000 Kishbek roubles (Kr).

The price of one 4″ wafer =

$$\frac{\textit{Total revenue}}{\textit{Total sales}} = \frac{10{,}000}{500} = 20$$

One 4″ wafer costs 20 Kr and therefore 100 4″ wafers = 20 × 100 Kr = 2000 Kr.

Q2 b) 100%

Sales of 6″ wafers in May = 200.

Sales of 6″ wafers in June = 400.

Recognize that if you double a number, you increase it by 100%. If this is not obvious, you can recall the formula for percentage change:

$$\%\ \textit{change} = \frac{\textit{actual change}}{\textit{original}} \times 100\% = \frac{200}{400} \times 100\% = 100\%$$

Q3 b) $156.25

First find the cost of one 6″ wafer. Five 6″ wafers cost 125 Kr, so one 6″ wafer costs $^{125}/_{5}$ = 25 Kr.

In May, 250 6″ wafers cost 250 × 25 Kr = 6,250 Kr. In May, the exchange rate = 40 Kr : 1 US $, so divide $^{6250}/_{40}$ to find the $ price = $156.25.

In August, the exchange rate = 20 Kr : 1 US $, so divide $^{6250}/_{20}$ to find the $ price = $312.50.

The difference = $312.50 – $156.25 = $156.25.

Section 15 Energy tariffs

Q1 e) £40

There are two factors to consider in the calculation: (1) the cost of energy and (2) the daily service charge.

Energy consumed (2,854 kwh) × unit price (1.32p) = approximately 3,700p.

+ consumption period (20 days) × daily charge (9.99p) = approximately 200p.

3,700p + 200p = 3,900p or approximately £40.

Q2 a) £45

The price of electricity on tariff B = 102.5% × the price of electricity on tariff C:

$$5.02\text{p} \times \frac{102.5}{100} = \textit{Tariff B electricity} \approx 5.15\text{p}$$

Energy consumed (640kwh) × unit price (5.15p) = approximately 3,300p

+ consumption period (91 days) × daily charge (13.39p) = approximately 1,200p

3,300p + 1,200p = 4,500p or £45.

Q3 c) £7.50

To find total for energy on tariff A and tariff B:

	(Energy consumed × unit price)	+	(Consumption period × daily charge)	
A =	(250kwh × 1.32p)	+	(15 days × 9.99p)	= 330p + 149.85p
B =	(120kwh × 5.15p)	+	(15 days × 13.39p)	= 618p + 200.85p

Total for energy on tariff A and tariff B = 1,299p or £12.99.
To find total for energy on tariff C:
Energy consumed (250 kwh) × unit price (1.25p) + (120 kwh) × unit price (5.02p) + annual service charge (£11.50) = 915p + £11.50.
Total for energy on tariff C = £20.65.
The difference in price = £20.65 – £12.99 = £7.66.
The question asks for the approximate difference, so choose the answer closest to £7.66.

Glossary

Terms

Arithmetic mean: The amount obtained by adding two or more numbers and dividing by the number of terms.

Compound interest: The charge calculated on the sum loaned plus any interest accrued in previous periods.

Denominator: The number below the line in a vulgar fraction.

Digit: One of the numbers 0,1,2,3,4,5,6,7,8,9.

Dividend: The number to be divided.

Divisor: The number by which another is divided.

Equivalent fractions: Two or more fractions with the same value.

Factor: The positive integers by which an integer is evenly divisible.

Fraction: A part of a whole number.

Fraction bar: The line that separates the numerator and denominator in a *vulgar fraction*.

Improper fraction: A fraction in which the *numerator* is greater than or equal to the *denominator*.

Integer: A whole number without decimal or fraction parts.

Interest: See 'Simple interest' and 'Compound interest'.

Lowest common denominator: The smallest common multiple of the denominators of two or more fractions.

Lowest common multiple: The least quantity that is a multiple of two or more given values.

Mean: See 'Arithmetic mean'.

Median: The middle number in a range of numbers when the set is arranged in ascending or descending order.

Mixed fractions: A fraction consisting of an integer and a fraction.

Mode: The most popular value in a set of numbers.

Multiple: A number that divides into another without a remainder.

Multiplier: A quantity by which a given number is multiplied.

Numerator: The number above the line in a *vulgar fraction*.

Prime factor: The factors of an *integer* that are *prime numbers*.

Prime factorization: The expression of a number as the product of its *prime numbers*.

Prime number: A number divisible only by itself and 1.

Proper fraction: A *fraction* less than 1, where the *numerator* is less than the *denominator*.

Proportion: Equality of *ratios* between two pairs of quantities.

Ratio: The comparison between two or more quantities.

Simple interest: The charge calculated on a loaned sum.

Vulgar fraction: A fraction expressed by numerator and denominator, rather than decimally.

Formulae used in this book

Chapter 1

$$Arithmetic\ mean = \frac{Sum\ of\ values}{Number\ of\ values}$$

$$Number\ of\ values = \frac{Sum\ of\ values}{Arithmetic\ mean}$$

$$Sum\ of\ values = Arithmetic\ mean \times Number\ of\ values$$

Chapter 3

Rates formulae

$$Distance = Rate \times Time$$

$$Rate = \frac{Distance}{Time}$$

$$Time = \frac{Distance}{Rate}$$

$$Average\ rate = \frac{Total\ distance}{Total\ time}$$

Work rate formula

$$Time\ combined = \frac{xy}{x + y}$$

$$\frac{1}{T1} + \frac{1}{T2} + \frac{1}{T3} = \frac{1}{T}$$

Chapter 4

Percentages formulae

$$Part = Percentage \times Whole$$

$$Whole = \frac{Part}{Percentage}$$

$$Percentage = \frac{Part}{Whole}$$

Percentage increase formula

$$\%\ increase = \frac{Actual\ amount\ of\ increase}{original\ whole} \times 100\%$$

$$New\ value = Original\ whole + Amount\ of\ increase$$

Percentage decrease formula

$$\%\ decrease = \frac{Actual\ amount\ of\ decrease}{original\ whole} \times 100\%$$

$$New\ value = Original\ whole - Amount\ of\ decrease$$

Simple interest

I = PRT

where I = Interest, P = Principal sum, R = Interest rate and T = Time period.

Compound interest

$I = P(1 + R)^{n-1}$

where P = the Principal sum, R = the Rate of interest and n = the Number of periods for which interest is calculated.

Chapter 5

$$Ratio = \frac{\text{'of' } x}{\text{'to' } y}$$

Recommendations for further practice

Useful websites

The following websites contain practice material for you to review or download. Web addresses do change from time to time and although this information is correct at the time of publication, currency of address information cannot be guaranteed.

http://www.shlgroup.co.uk
SHL is a leader in test preparation. SHL tests are a commonly used test for graduate roles. The website contains practice for verbal, numerical and diagrammatical reasoning tests.

http://www.work.guardian.co.uk/psychometrics
Practice tests useful for tests such as those produced by SHL.

http://www.barcap.com/graduatecareers/barcap_test.pdf
Barclays Capital website contains practice tests for the SHL tests, including diagrammatic reasoning, verbal reasoning and numerical reasoning.

http://www.morrisby.com
Morrisby are leaders in psychometric testing and offer advice and support for test-takers. The website contains information on test-taking and practice questions.

http://www.ets.org
Educational Testing Service, the US leading testing company. ETS specializes in the preparation of the SAT, GRE, GMAT and LSAT aptitude tests. The Web site contains specialist information on test preparation and provides practice test for US standardized tests.

http://www.deloitte.co.uk/index.asp
Deloitte and Touche are a leading tax and advisory consultancy. The website contains practice tests for accountants and non-accountants.

http://www.pgcareers.com/apply/how/recruitment.asp
Proctor and Gamble's website contains problem-solving tests containing 50 questions, testing verbal reasoning, data interpretation and numerical reasoning questions.

http://www.pwcglobal.com/uk/eng/carinexp/undergrad/quiz.html
PriceWaterhouseCoopers' website contains multiple choice questions similar to those used in the PWC graduate tests.

http://www.thewizardofodds.com/math/group1.html
A compendium of logic puzzles requiring mathematical reasoning skills.

http://www.home.g2a.net
Logic problems requiring mathematical reasoning skills to arrive at the solutions. The problems are accompanied by explanations.

http://rinkworks.com/brainfood/p/eg1.shtml
Mathematical reasoning puzzles, which require both logical and mathematical reasoning skills.

http://www.publicjobs.gov.ie/numericatest.asp
The Civil Service Commission website contains online practice tests and explanations.

http://www.resourceassociates.com/html/math.htm
Resource Associates is a US-based human resources consultancy. The website contains practice data interpretation tests.

Further reading from Kogan Page

The Advanced Numeracy Test Workbook
ISBN 978 0 7494 3941 5

Aptitude, Personality & Motivation Tests
ISBN 978 0 7494 4179 1

The Aptitude Test Workbook
ISBN 978 0 7494 3788 6

A-Z of Careers & Jobs
ISBN 978 0 7494 4977 3

Career, Aptitude & Selection Tests
ISBN 978 0 7494 4819 6

Graduate Psychometric Test Workbook
ISBN 978 0 7494 4331 3

Great Answers to Tough Interview Questions
ISBN 978 0 7494 4356 6

How to Master Psychometric Tests
ISBN 978 0 7494 4279 8

How to Pass Advanced Aptitude Tests
ISBN 978 0 7494 3787 9

How to Pass Advanced Numeracy Tests
ISBN 978 0 7494 3791 6

How to Pass Advanced Verbal Reasoning Tests
ISBN 978 0 7494 4969 8

How to Pass the Civil Service Qualifying Tests
ISBN 978 0 7494 4853 0

How to Pass the GMAT
ISBN 978 0 7494 4459 4

How to Pass Graduate Psychometric Tests
ISBN 978 0 7494 4852 3

How to Pass the New Police Selection System
ISBN 978 0 7494 4946 9

How to Pass Numeracy Tests
ISBN 978 0 7494 4664 2

How to Pass Numerical Reasoning Tests
ISBN 978 0 7494 4796 0

How to Pass Professional Level Psychometric Tests
ISBN 978 0 7494 4207 1

How to Pass Selection Tests
ISBN 978 0 7494 4374 0

How to Pass Verbal Reasoning Tests
ISBN 978 0 7494 4666 6

How to Succeed at an Assessment Centre
ISBN 978 0 7494 4421 1

IQ and Aptitude Tests
ISBN 978 0 7494 4931 5

IQ and Personality Tests
ISBN 978 0 7494 4954 4

IQ and Psychometric Tests
ISBN 978 0 7494 5106 6

IQ and Psychometric Test Workbook
ISBN 978 0 7494 4378 8

The Numeracy Test Workbook
ISBN 978 0 7494 4045 9

Preparing the Perfect Job Application
ISBN 978 0 7494 5022 9

Preparing the Perfect CV
ISBN 978 0 7494 4855 4

Preparing the Perfect CV
ISBN 978 0 7494 4855 4

Readymade CVs
ISBN 978 0 7494 4274 3

Successful Interview Skills
ISBN 978 0 7494 4508 9

Test Your IQ
ISBN 978 0 7494 4833 2

Test Your Numerical Aptitude
ISBN 978 0 7494 5064 9

Test Your Own Aptitude
ISBN 978 0 7494 3887 6

The Ultimate CV Book
ISBN 978 0 7494 3875 3

The Ultimate Interview Book
ISBN 978 0 7494 4310 8

The Ultimate IQ Test Book
ISBN 978 0 7494 4947 6

The Ultimate Job Search Book
ISBN 978 0 7494 4690 1

The Ultimate Psychometric Test Book
ISBN 978 0 7494 4458 7

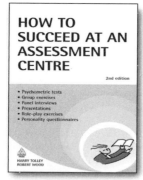

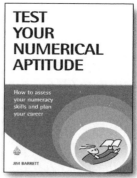

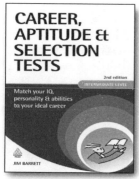

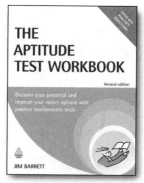